AMERICA'S SPACEPORT
CAPE CANAVERAL

DONALD D. SPENCER

Schiffer
Publishing Ltd

4880 Lower Valley Road, Atglen, Pennsylvania 19310

Other Schiffer Books by the Donald D. Spencer:

A History of the Alligator: Florida's Favorite Reptile, 978-0-7643-3083-4, $24.99
Greetings from Orland & Winter Park, Florida: 1902-1950, 978-0-7643-2966-1, $14.99
St. Augustine, Florida: Past and Present, 978-0-7643-3146-6, $24.99
Greetings from St. Augustine, 978-0-7643-2802-2, $24.95
Greetings from Tampa, Florida, 978-0-7643-2898-5, $24.95
Greetings from Sarasota, Florida: Bradenton & Surrounding Communities, 978-0-7643-3213-5, $24.99
St. John's River: An Illustrated History, 978-0-7643-2826-8, $24.99
Greetings from Jacksonville, Florida, 978-0-7643-2958-6, $24.99
Greetings from Ormond Beach, Florida, 978-0-7643-2809-1, $24.95
Daytona Bike Week, 978-0-7643-2977-7, $24.99
Greetings from Daytona Beach, 978-0-7643-2806-0, $24.95
Greetings from Palm Beach, Florida: 1900s-1960s, 978-0-7643-3263-0, $24.99
'50s Roadside Florida, 978-0-7643-3364-4, $24.99
Greetings from Fort Myers & Sanibel Island, 978-0-7643-3305-7, $24.99
Mid-Century Vegas: 1930s-1960s, 978-0-7643-3129-9, $39.99

Schiffer Books are available at special discounts for bulk purchases for sales promotions or premiums. Special editions, including personalized covers, corporate imprints, and excerpts can be created in large quantities for special needs. For more information contact:

Schiffer Publishing Ltd.
4880 Lower Valley Road
Atglen, PA 19310
Phone: (610) 593-1777; Fax: (610) 593-2002
E-mail: Info@schifferbooks.com

For the largest selection of fine reference books on this and related subjects, please visit our web site at **www.schifferbooks.com** We are always looking for people to write books on new and related subjects. If you have an idea for a book please contact us at the above address.

This book may be purchased from the publisher. Include $5.00 for shipping. Please try your bookstore first. You may write for a free catalog.

In Europe, Schiffer books are distributed by
Bushwood Books
6 Marksbury Ave.
Kew Gardens
Surrey TW9 4JF England
Phone: 44 (0) 20 8392 8585; Fax: 44 (0) 20 8392 9876
E-mail: info@bushwoodbooks.co.uk
Website: www.bushwoodbooks.co.uk

Designed by Stephanie Daugherty
Type set in Serpentine/ NewBskvll BT/ Humanst521 BT

ISBN: 978-0-7643-3616-4
Printed in China

DEDICATION

To the seven U.S. astronauts of the Space Shuttle Challenger who lost their lives January 28, 1986: Gregory Jarvis, Sharon Christa McAuliff, Ronald E. McNair, Ellison S. Onizuka, Judith Resnik, Francis R. (Dick) Scobee, and Michael J. Smith, to whom a nation is eternally grateful.

ACKNOWLEDGMENTS

I would like to give special thanks to the dedicated men and women of NASA for their ongoing commitment and perseverance in achieving what many consider to be not achievable. The American people can be proud of them and proud of their Space program, as NASA continues its ongoing mission to pioneer a new and exciting future.

I would also like to thank NASA for the wonderful images from Space and their willingness to share these images with the general public. Special thanks to Margaret Persinger at the Kennedy Space Center; Deborah Stillwell at the Patrick Air Force Base, and especially Jodi Russell, of the Johnson Space Center, for providing me with thousands of NASA images.

I would also like to thank my son Steven Spencer for the loan of his Space book library and Space memorabilia collection.

This book is based mainly on the many wonderful images from the National Aeronautics and Space Administration (Johnson Space Center, Kennedy Space Center, NASA Headquarters, George C. Marshall Flight Center) and the postcard and photo collection of the author augmented by illustrations and photographs from the author's son, Steven, pages 34, 35, 38, 39, 40, 62 (bottom), 90 (bottom center); Dover Publications, Inc., pages 4 (left), 36 (right), 37 (right); and United States Air Force, pages 20 (left), 26 (right center), 26 (bottom center), 26 (bottom left), 27(bottom center), 27 (bottom right), and 27 (top left).

CONTENTS

PREFACE

THE STORY OF OUR EXPLORATION OF SPACE IS A SHORT ONE, BEGINNING IN OCTOBER 1957 WHEN the Soviet Union launched *Sputnik 1*, the world's first artificial satellite. Since this time we have witnessed many spectacular feats of Space exploration. Most of these would have been considered figures of fancy not so many years before when the phrase "Reach for the Moon" meant "striving for the impossible."

More than fifty years later, though, twelve men have walked on the Moon and brought back moon rock for scientists on Earth to study. Unmanned Space probes have landed on the surface of the nearest planets, Mars and Venus, and sent back pictures of the local scenery. Navigation, Earth Resources, Communications, and Meteorological satellites have become reliable workhorses in the service of mankind. *The International Space Station* has resulted in many nations working together for a common goal. Space probes have detected water on the Moon.

We are already living in the Space Age, yet we have only just begun to leave the surface of our own planet. The commercialization and industrialization of Space will continue in the fields of communications, weather forecasting, and others. A commercial suborbital Space vehicle is expected to be operational in the not to distant future. Space probes have already been launched that will further explore the planets in our Solar System. One day a spacecraft may leave Cape Canaveral and carry astronauts to Mars. We can be sure that there will be no turning back now that the foundations of our journey into Space have been laid.

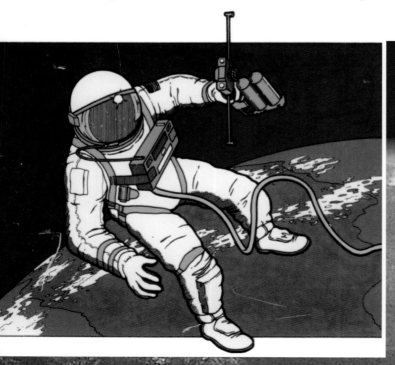

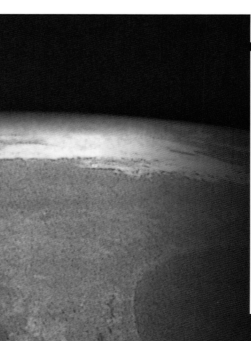

Introduction:
FROM LAND TO AIR

CANAVERAL IS A SPANISH WORD MEANING "A PLACE OF ROOTS OR CANE." THE SPANISH named the cape Canaveral because there were cane or reeds growing here. Ponce de Leon called the point The Cape of Currents, but the name Canaveral appeared on the earliest Spanish maps of Florida. A few years ago it was just a jut of sand in a world of dunes, a small elbow of palmetto scrubland, and a narrow ledge of beach that protruded out into the Atlantic Ocean. Miles from the mainland, miles from anywhere, it was a drowsy place, a sleepy world of a quiet and whispering sea, sunny days and starlit nights, and lazy Sundays spent hunting and fishing under the warm Florida sun. A few small farms dotted its sandy soil. This sparsely populated, triangular-shaped piece of land covered roughly 15,000 acres. This was Cape Canaveral.

A MISSILE SHOOTING GALLERY

Nothing stands still for long here on our Earth. Time moves steadily in the pulse of ages. All things change; it is inevitable in the slow march of years and centuries. People are born, they fight, and die; more people follow them, more people fight, and new killing tools are invented to do more fighting. The caveman's club gave way to the spear and arrow, these to the musket and saber, these to the rifle and heavy artillery, to even heavier artillery, and now to rockets, missiles, and satellites.

This has started what the world now has learned to call and know as Cape Canaveral. Men came to The Cape, looked it over, and decreed it must be made into a military base for the testing of missiles — and so the sand dunes and palmetto scrub land were smoothed, the marshes filled in, the farms and orange groves pushed away, and the quiet sleepy area of Cape Canaveral vanished in a roaring bedlam of men and machines and hurried, massive construction.

The Cape became the Cape Canaveral Missile Testing Center, a maze of blockhouses and gantry towers, of steel and concrete, of tall, sleek missiles

Satellite View of Florida. This view from a satellite in space clearly shows the Cape on Florida's east coast jutting into the Atlantic Ocean.

Kennedy Space Center, Florida. These are the Launch Pads where many of the U.S. Air Force missiles, unmanned rockets carrying NASA space probes, and early NASA flights were sent into space. Launch Complex 14 is where John Glenn lifted-off aboard the *Friendship* 7 space capsule in 1962; he became the first American astronaut to orbit the Earth. Launch Complex 19 is where ten two-manned Gemini flights were launched using modified *Titan 2* missiles to test the maneuvers and hardware required for a Moon landing. Launch Complex 34 is where the launch of the first manned orbital test flight of the Apollo spacecraft, *Apollo 7*, took place in 1968 atop a *Saturn 1B* rocket with three astronauts aboard. Pad 34 is also the site where three astronauts perished when their *Apollo 1* spacecraft caught fire during a countdown test in 1967.

resting on their launching pads wreathed in vapor, and of restless, ceaseless activity day and night. Its control centers and instrument stations have mushroomed, thick-walled and sturdy. Almost overnight it became a teeming base that clattered and roared with the thunder of creation — a strange land where giant missiles shook the Earth and only scientists dreamed and predicted the future. The Cape was now a 5,000-mile missile shooting gallery of open water.

In the beginning only the rapid construction of the missile center was spectacular. Then came men shooting skyrockets into the sky. After a while, they became larger and more powerful and, before long, there was talk of sending a missile — a satellite — to orbit the Earth. It was almost like a Jules Verne story coming true. Missiles thundered from their launching pads in bursts of flames that lit up the entire area and the rockets went up on their designated paths, but no satellites went into orbit... America's satellites remained in the heads of our scientists and mathematical equations.

MAN IN SPACE

History changed on October 4, 1957, when the Soviet Union launched the world's first satellite into space. The success of this primitive satellite initiated the "space race" that would be the leading preoccupation for the world's two most powerful nations for the following decade. In 1958, the United States formed the National Aeronautics and Space Administration, now known throughout the world as NASA. From its inception, people from all over the country united for a common goal: the exploration and research of the last frontier — Space.

Another important event happened in 1958 — and it occurred at Cape Canaveral. America launched its first satellite into Earth orbit. With the launch of *Explorer 1*, a completely new technology was born, a technology that harnessed the computer for totally new applications — engineering, planning, design, coordination, administration, real-time quality and operational control, and decision-making—all with previously unheard of speed, efficiency, and precision.

The Race to Space. On October 4, 1957, the Soviet Union's *Sputnik 1* satellite could be seen clearly in the night sky. At about twenty-three inches wide and 185 pounds, it glided overhead like a shooting star. In the following month the Soviet Union launched the 1,120-pound *Sputnik 2* that carried a dog named Laika. America didn't have a successful space launch until January 31, 1958, when *Explorer 1* became the country's first artificial satellite to circle the globe. On March 17, 1958, the U.S. launched *Vanguard 1* from Cape Canaveral. *Vanguard 2*, a 22-pound satellite, was launched February 17, 1959. The National Aeronautics and Space Administration (NASA) opened for business October 1, 1958, and Cape Canaveral was instantly transformed from scrubland into a major missile-launching center. The race to space had begun and it became a race between the United States and the Soviet Union, competing to achieve the next historic space "first." We have gone from simple satellites like the *Sputnik, Explorer,* and *Vanguard* to complex communications satellites that now circle the globe, providing a network of worldwide communication. We have witnessed the journey of robot probes to faraway planets, landed American astronauts on the Moon, and built an *International Space Station*, complete with shuttling vehicles that carry astronauts into space. This magnificent view is where Earth appears like a blue-and-white marble, unlike any other planet, in this image built up from many photos taken by NASA spacecraft.

KENNEDY SPACE CENTER

The area began to change character when NASA established its own facilities on The Cape and adjacent Merritt Island. The Kennedy Space Center became the doorway to outer space. From here, the United States became engaged in a long-term, multifaceted program to explore and study our solar system, reaching out to the moon, the sun, and the planets. The initial projects have already resulted in many practical applications of space technology, though the surface has barely been scratched. The future is just beginning to unfold here in Florida.

The Kennedy Space Center at Cape Canaveral is NASA's leading space launch operation. In its role as chief developer of launch procedures, technology, and facilities, the Kennedy Space Center launches manned space vehicles, unmanned planetary spacecraft, and scientific, meteorological, and communication satellites.

The Center's famed Launch Complex 39 served as the launch site for American astronauts' journeys to explore the moon, to occupy the orbiting *Skylab* space station, and to dock with a Soviet *Soyuz* spacecraft on the first joint U.S.-U.S.S.R. manned mission. With these and other programs successfully completed, Launch Complex 39 was modified to accommodate the Space Shuttle, the world's first reusable spaceship, destined to become the space transportation system from the 1980s to early 2011.

PROJECTS MERCURY & GEMINI

While the landing of American astronauts on the Moon was an early major goal of NASA, the three projects directed toward that stunning achievement — Projects Mercury, Gemini, and Apollo — were designed as well to lay the foundations for future space programs of even greater challenge and promise.

Project Mercury, the initial step in the American manned space flight program, was organized in 1958 and completed in May 1963, after six successful manned missions, with astronaut Alan B. Shepard, Jr. becom-

Kennedy Space Center. The 140,000-acre Kennedy Space Center (KSC) is NASA's major spaceport, located on the east coast of Florida on Merritt Island. Next to it is the Cape Canaveral Air Force Station, launch site for the Air Force Eastern Space and Missile Test Range, which extends southeastward across the Atlantic Ocean. KSC itself has only two launch pads, 39A and 39B. These were used for all but one of the Apollo launches and for all the Space Shuttle launches. All expendable rocket launches and all pre-Apollo manned launches were from the Air Force Station, where most of the launch pads have now been deactivated. Here, the Space Shuttle *Endeavor* is leaving Launch Pad 39B. There is a 15,000-foot runway at KSC, which the staff checks for stray alligators and bobcats prior to a Space Orbiter landing.

HISTORY OF MANNED SPACE PROGRAMS

MERCURY 1958-1963

GEMINI 1965-1966

APOLLO 1968-1972

SKYLAB 1973-1974

SHUTTLE 1981-PRESENT

ing America's first man in space in the little one-man Mercury spacecraft and astronaut John H. Glenn, Jr. becoming the first American to orbit the Earth.

Project Gemini was the intermediate step toward achieving a manned lunar landing, bridging the gap between the short-duration Mercury flights and the long-duration Apollo missions. The Gemini program, involving a larger and heavier two-man spacecraft, was completed in November 1966, following ten Earth-orbital missions. Edward H. White, II, became America's first astronaut to "walk" in space.

PROJECT APOLLO

People have dreamed of exploring the Moon for hundreds of years. The United States and Russia made the dream a reality in the mid-twentieth century. In 1959, *Luna 1*, the first spacecraft to leave Earth's gravity, was launched by the Soviet Union toward the Moon. A decade of intense space activity followed as Soviet Union and American probes, robots, and manned crafts were sent to investigate and land on the lunar surface. *Luna 2* became the first lunar probe to hit the Moon when it

crash-landed on the surface in 1959. The next month, *Luna 3* took the first photographs of the far side. The U.S. also sent *Ranger* and *Surveyor* probes to the Moon. *Ranger 7* crashed on the Moon in 1964 and returned the first close-up images, taking 4,308 photographs. *Apollo 8* carried the first astronauts around the Moon in 1968, making ten orbits.

The following year, on July 2, 1969, *Apollo 11* astronauts executed history's first lunar landing from their huge, three-module craft, and astronauts Neil A. Armstrong and Edwin E. "Buzz" Aldrin, Jr. emerged from the Lunar Module and set foot upon an area known as the Sea of Tranquility while astronaut Michael Collins remained in the Command Module. After the 195-hour mission, craft and crew splashed down in the Pacific Ocean on July 24. The outstandingly successful Apollo program ended in December 1972, with *Apollo 17* fulfilling the final lunar mission. The primary purpose of Project Apollo was to land humans on the Moon and return them to Earth. Of the eleven Apollo missions, six — *Apollos 11, 12, 14, 15, 16 and 17* — met this objective. Twelve people have now walked on the Moon. The astronauts of *Apollo 15, 16,* and *17* also drove a jeep-like electric car, called the Lunar Rover, on the Moon.

Man on the Moon. Not before or since has so much attention been focused on a space mission. *Apollo 11*, following two manned flights in Earth orbit and two manned flights in lunar orbit, would fulfill an age-old dream of mankind. The fateful day was July 16, 1969 — the huge *Saturn V* rocket carrying *Apollo 11* rose gracefully into the air from the Kennedy Space Center. The crew, whose names would go down in history, were experienced astronauts from the Gemini program: Neil Armstrong, Edwin Aldrin, and Michael Collins. On July 20, the Lunar Module *Eagle* separated from the Command Module *Columbia* on the far side of the Moon. *Eagle*, with astronauts Armstrong and Aldrin, landed on the Moon. The two astronauts collected samples, planted the American flag, and conducted experiments during a two-hour and thirty-one minute stay. The *Eagle* blasted off, docked with Command Module *Columbia,* and splashed down in the Pacific Ocean on July 24.

SKYLAB

Following the Apollo Moon Program, NASA initiated an eight-month mission toward two major objectives: 1. Prove humans could live and work in space for extended periods; and 2. Gain a greater understanding of solar astronomy.

Skylab, the United States' first experimental space station, used much of the hardware and technology developed for previous manned programs. On May 14, 1973, an unmanned Saturn V rocket launched *Skylab* into a near-circular orbit 270 miles above the Earth's surface. Three successive crews orbiting in Apollo spacecraft rendezvoused and docked with the space station, working in it for a total of 171 days, during which time *Skylab* traveled more than 70 million miles in orbit around the Earth. Scientists say that the data created by approximately three hundred scientific and technical experiments conducted on *Skylab* have contributed greatly to our understanding of solar physics and potential uses of the Sun's vast energy.

APOLLO-SOYUZ TEST PROJECT

The first international manned spaceflight, Apollo-Soyuz, tested the compatibility of rendezvous and docking systems between American and Soviet spacecraft. Aboard the docked spacecraft, the astronauts would speak to the cosmonauts in Russian and the cosmonauts would reply in English.

Seven hours after the *Soyuz* spacecraft fired into orbit from its launch pad at Kazakhstan, NASA launched the last *Apollo* spacecraft from Launch Pad 29B. Two days after the July 15, 1975 launch, *Apollo* docked with *Soyuz*. The Apollo-Soyuz project was designed to give engineers on both sides the experience of collaborating to build hardware that was mutually compatible and to test it in flight. Over the following two days, the *Apollo* astronauts and *Soyuz* cosmonauts conducted various experiments. This historic mission marked the way for future joint manned flights and, in the event of an emergency, proved international space rescue missions were possible.

THE SPACE SHUTTLE — A REUSABLE SPACECRAFT

Man at work in Space — to be exact, inside an aerospace vehicle that takes off like a rocket, maneuvers in orbit like a spacecraft, and lands like an airplane. Man can now walk in space without a lifeline, depending on a nitrogen-propelled, hand-controlled backpack device called the manned maneuvering unit (MMU).

The first test flight of the Space Transportation System (STS) was in 1981. STS, usually referred to as the Space Shuttle, is made up of an Orbiter with three main engines, an external tank, and two solid rocket boosters. Cargo is carried to space in the Orbiter's payload bay. Propellant for the main engines is supplied from the external tank. After each mission, the Orbiter returns to Earth, gliding to a landing on a very long runway. The STS launches satellites and probes, carried the *Spacelab* space station and the *Hubble Space Telescope*, and provides a platform for construction and repairs in space. Each Orbiter can carry a crew of seven and stay in orbit for at least ten days. The Orbiter's cabins have

Space Shuttle on its Way to the International Space Station. *International Space Station* assembly flights began in 1998 and dominated operations at Kennedy Space Center thereafter, as the Space Shuttle blasted into space again and again, every few months. The station is made up of numerous pieces small enough to fit in the Shuttle's cargo bay, each of which are joined to the station via astronaut and cosmonaut spacewalk work — construction work at an altitude of 220-miles above the Earth.

Space Shuttle Launch at KSC. The Space Shuttle is made up of a reusable orbiter, the expendable external tank, and two reusable solid-fuel rocket boosters. Takeoff is vertical; the tank and boosters are jettisoned during the ascent, after which the orbiter is left on its own. It then reenters the atmosphere and lands — *un-powered* — in the manner of a glider. The shuttles were given attractive names—*Atlantis, Discovery, Endeavor, Challenger,* and *Columbia*, but although many successful flights were made, the *Challenger* was destroyed during its launch in 1986 and the *Columbia* broke up during reentry in 2003.

three decks — the flight deck, mid-deck, and a lower deck that houses life-support equipment.

The knowledge and skills acquired in the Mercury, Gemini, and Apollo programs are today being applied to the Space Shuttle program, which opens an entire new era of space operations. The Shuttle is designed to reduce the cost and increase the effectiveness of using space for commercial, scientific, and defense needs. Space Shuttle *Columbia* made the first flight into space. Launched from the Kennedy Space Center on April 12,

1981, *Columbia* landed fifty-four hours later at Edwards Air Force Base in California. This space mission opened a new era when astronauts John Young and Robert Crippen demonstrated the "space-worthiness" of the first Space Shuttle.

Unlike the spacecraft of the past, the Space Shuttle is a true spaceship — a reusable vehicle affording routine access to space. The Shuttle is expected to remain operational into 2011. Approximately as large as a DC-9 jetliner, the Shuttle has a cargo compartment sixty feet long and fifteen feet in diameter, which can accommodate a wide variety of payloads, such as the complicated but reusable Spacelab, the space-borne laboratory built by the European Space Agency that was successfully put to the test in the Space Shuttle for the first time in 1983.

The Space Shuttle is the world's most reliable and versatile launch system. Depending upon the mission's objectives, crew sizes aboard the Orbiter usually vary between five and seven members. Transporting cargo into orbit 100 to 217 nautical miles above the Earth is the program's primary purpose. In order to deliver different types of equipment, spacecraft, and scientific experiments, the Shuttle can be configured to accommodate its cargo, or payload. The cargo bay can carry a payload up to 65,000 pounds.

Due to its ability to make routine scheduled flights, the Space Shuttle has developed into an important commodity for the space program. The Shuttle's versatility has made it possible to repair costly satellites in space, have astronauts deploy satellites from the Shuttle's cargo bay, deliver large payloads, and take supplies to the International Space Station.

MIR-SPACE SHUTTLE COOPERATIVE PROJECT

The idea of living and working in space has fascinated scientists and engineers for a long time.

On February 19, 1986, the Soviet Union launched the *Mir Space Station*, designed to be more substantial than the earlier American *Skylab* and Soviet *Salyuts* space stations. The goal was to man the station for several years.

The *Mir Space Station* was built by connecting several Mir modules. The Core module provided the living quarters for the cosmonauts. Added to the Core module were several additional modules: Kvant I (astronomical research), Kvant 2 (improved life support systems), Kristall (material processing laboratories), Spektr (general experiments involving American astronauts), and Priorda (remote sensing).

After the disintegration of the Soviet Union, joint missions were agreed upon between the United States and Russia. In the 1990s a docking module was attached to Mir so that American Space Shuttles could link up with it. The Russians had previously relied upon their much less elaborate Proton servicing rockets. The docking module was taken up by the Space Shuttle *Atlantic* on November 12, 1995. The *Mir Station* gave American astronauts practice in space docking and space station experience at a time when America didn't have their own facility in space.

Mir provided living and working quarters. A series of Shuttle missions transported American astronauts to Mir where they served as members of the crew. Before reaching Mir, American astronauts received specialized training and cosmonaut certification, as well as learning the Russian language.

Mir's interior was described as a general store, crowed with scientific instruments, cables and hoses, as well as everyday items. Before August 1999, Mir was continuously manned, apart from two short breaks, with three Soviet cosmonauts or NASA astronauts. Occasionally though there were more than three occupants — and some of these occupants were researchers from other nations.

The Mir-Space Shuttle cooperative project brought NASA a great deal further in working with an international partner on space operations. Many of the Mir scientific experiments had great value, especially those in medical research, biology, and chemistry.

In the late 1990s, Mir had several problems and had to be destroyed. The end came March 23, 2001, when it broke up in the Earth's atmosphere; fragments fell into the sea between New Zealand and Chili. In all, it stayed up for 5,511 days and completed just over 89,000 orbits at a height of 240 miles above the Earth.

INTERNATIONAL SPACE STATION

The United States and Russia launched the first parts of the *International Space Station* (ISS) in 1998; Brazil, Canada, the European Space Agency, and Japan have also contributed. The station is made up of more than one hundred elements. The biggest contributions — including a connecting node, solar panels, habitation module, an unpressurized module, and several laboratories — are from the United States. A Core module, providing living quarters for the first few years, comes from Russia. Canada provided a robot arm. Most of the participating space agencies will help transport supplies to the station.

After 1998, Space Station assembly flights dominated launch operations at Kennedy Space Center. Again and again, every few months, the Space Shuttle blasted into space from Launch Pad 39 carrying new modules to add to the station. The station is made up of numerous pieces small enough to fit into the Space Shuttle's cargo bay, each of which would be integrated into the Space Station. A power module, solar panels, a crane, and a dwelling module, piece by piece, were joined together in space. The final design, with outstretched solar panels, is longer than a football field and weighs nearly 550 tons.

The *International Space Station*, designed to serve as a micro-gravity laboratory, carries a permanent crew. Astronauts and cosmonauts from many nations carry out a wide-ranging program of research and experiments in biology, physics, and materials science.

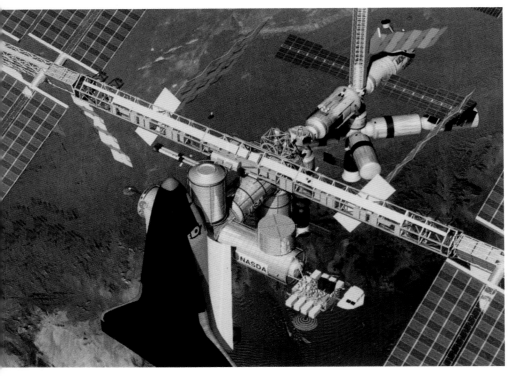

International Space Station (ISS). The largest structure ever to fly in space, the *International Space Station* (ISS) program involves several nations and is controlled by the U.S. and Russia equally. It is permanently occupied; resident crews usually stay in orbit for six months. Supplies are delivered by America's Space Shuttle and Russia's *Progress* cargo ship. The ISS will cover an area as big as two football fields and weigh 475 tons when it is fully assembled in space in 2010 and people on Earth will be able to track it in orbit with the naked eye. It will be one of the brightest objects in the sky, brighter than most stars. This view shows the Space Shuttle docked at the ISS.

HUBBLE SPACE TELESCOPE

The ultimate telescope for astronomers seeking pin-sharp views of the depth of the universe is the *Hubble Space Telescope*. Conceived during the 1940s, the *Hubble Space Telescope* was designed and constructed in the 1970s and 1980s. The crew of the Space Shuttle *Discovery* deployed the 2.4-meter reflecting telescope in low-Earth orbit on April 25, 1990. The *Hubble Space Telescope* is a cooperative program of NASA and the European Space Agency with the goal of operating a long-lived space-based observatory that will benefit the international astronomical community. Equipped with three cameras, two spectrographs, and five guidance sen-

sors, the Hubble relays images to Earth, enabling scientists to study the far reaches of the Universe.

The Hubble is an unmanned observatory in orbit far above the clouds and atmosphere haze that block the view of Earth-based telescopes. Astronomers from dozens of countries use the Hubble, operating it by remote control. Sensitive light detectors have replaced the human observer at the eyepiece and electronic cameras record exquisite views of the cosmos.

SATELLITES

Anything in orbit around another object can be called a satellite. The Moon, for example, is a natural satellite of Earth. Since 1957, thousands of artificial satellites have been launched into orbit around Earth. They come in many shapes and sizes and occupy different types of orbits, depending on what their purpose is. Many communication satellites occupy geostationary orbit, for example, while many weather satellites are in polar orbit.

Communications Satellites

Telephone calls, television broadcasts, and the Internet can all be relayed by communication satellites. These satellites connect distant places and make it possible to communicate with remote areas. The world's first geostationary satellite for commercial communications was *Early Bird* (Intelsat I), launched at The Cape in 1965.

Meteorological Satellites

Meteorological satellites can see the way weather systems develop and move around the globe. They record images that are then broadcast daily on television, show cloud cover, and monitor hurricanes growing and moving across the oceans. Meteorological satellites also carry instruments to take readings, which are converted to the pressures, humidity, and temperatures that are needed for weather forecasting. The first weather satellites launched at The Cape were the *Tiros* in 1960 and *Nimbus 4* in 1970. Space flights have provided many spectacular photographs of our planet, and among the most beautiful are those showing the swirling masses of cloud that trace the pattern of weather systems across the face of the globe.

Navigation Satellites

To steer an accurate course between two places, a navigator needs to know his or her exact position. For thousands of years, sailors calculated their positions using the Moon, Sun, and Stars. Satellites transmit radio waves that can be detected on Earth even when clouds obscure the sky. As a result, navigation is now possible in any weather.

By the late 1990s, the Global Positioning System (GPS) had become the most reliable and accurate navigation system ever. GPS consists of twenty-four satellites as well as equipment on the ground. *Transit* was the first satellite navigation system. It was launched at The Cape in 1964 to improve position location of Polaris nuclear submarines. In 1978, the U.S. Air Force launched the first satellite that it acknowledged to be a GPS satellite. More GPS satellites were launched at The Cape in the 1980s and 1990s.

Earth Resources Satellites

Satellites that help scientists study the Earth's surface are called Earth Resources Satellites. They can show whether icecaps are melting or crops are failing and pinpoint resources such as coal or metal ores. This is possible because the satellites' instruments analyze light and other radiation reflected and emitted from surface features. Earth Research Satellites pass regularly over the whole globe, allowing scientists to produce maps that trace how a particular area changes over time. In 1972, *Landsat I* was launched and took the first combined visible and infrared image of Earth's surface. Six years later the *Seasat* satellite made the first valuable measurements of oceans with radar. In 1992, the *Topex/Poseidon* mission began collecting ocean data in unprecedented detail.

Military Satellites

Many of the earliest satellites were made for the Armed Services. Military satellites are widely used today to gather information about battlefields. They take pictures so detailed they can show where a person is standing, locate missing troops, and provide secure communications. Some satellites monitor the globe, looking for signs of a nuclear explosion or launch of a nuclear missile.

ROBOT EXPLORERS

Celestial objects, whether viewed unaided in the night sky or seen through telescopes and in pictures returned by spacecraft, are of compelling beauty. There are eight planets in our Solar System. The four innermost planets are solid, but only Earth — and possibly Earth's only natural satellite, the Moon — has water on its surface. The Earth is the only planet that has a life-supporting atmosphere. Mercury and Venus are closer to the Sun and too hot for life. Mars is farther away, but too cold.

Beyond these inner planets lies the asteroid belt, a zone of rocks and mini-planets ranging in size from a few feet to several hundred miles across. Farther out are the four gas giants — Jupiter, Saturn, Uranus, and Neptune — which are huge frozen worlds of hydrogen, helium, and other gases. Beyond all them is Pluto, a tiny ball of dust and ice spinning through space like a dirty snowball, almost four billion miles from the Sun.

Our only close-up views of Mercury date from the 1970s when the planetary probe *Mariner 10* flew by and revealed Mercury to be a heavily cratered world. Since the 1962 flyby of *Mariner 2*, many Russian (*Veneras* and *Vegas*) and U.S. (*Pioneer* and *Venus*) probes have sent back images of Venus.

Since 1965, the United States (*Mariner*, *Viking*, *Global Surveyor*, *Pathfinder*, *Climate Orbiter*, *Polar Lander*), Russia (*Mars 3*), and Japan (*Nozomi*) probes have investigated Mars in an effort to better understand the planet. Launched from Cape Canaveral in 2003, twin rovers *Spirit* and *Opportunity* landed to explore Mars. The 1973 *Pioneer 10*, the 1979 *Voyager 1*, and the 1995 *Galileo* probes studied the planet Jupiter. Four probes have flown past

Communications Satellite Launch. A British *Skynet* communications satellite was launched from the Cape November 21, 1969. The satellite was to enter a synchronous orbit 22,500 above the Earth over the Indian Ocean off the East Coast of Africa. It was the first of two satellites to be orbited to provide Great Britain with a secure communications link between points as far apart as England and Singapore.

Pioneer 10 Stamp. In the 1970s the USPS issued this 33-cent stamp illustrating the *Pioneer 10* spacecraft.

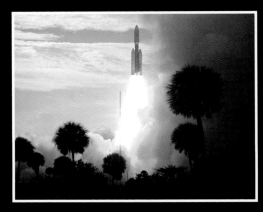

Viking 1 Spacecraft Leaves Earth to Visit Mars. The *Viking 1* spacecraft, launched August 20, 1975 from Cape Canaveral, arrived at Mars on June 19, 1976. The lander separated from the orbiter and touched down at Chryse Planitia on July 20. The primary mission of the spacecraft was to obtain high-resolution images of the Martian surface, characterize the composition and structure of the surface and atmosphere, and search for the evidence of life. The *Viking 1* lander and orbiter sent back thousands of images. The spacecraft's mission ended on August 7, 1980, after over 1,400 orbits. It found no clear evidence of living microorganisms.

Pioneer 10 Sailing Past Jupiter. On March 3, 1972 the *Pioneer 10* was the fastest spacecraft ever to fly, leaving Cape Canaveral at 32,107 miles per hour. It had a high-gain dish antenna, a configuration that was made necessary by the astronomical distances its transmissions must travel to Earth. Spindly outriggers held the nuclear generators that powered the craft. *Pioneer 10* was on its way to study and photograph planets in outer space. The first images of Jupiter were returned on November 6, 1973. Closest approach to the planet was on December 3, 1973, at a distance of 80,780 miles. More than three hundred images were returned. The Jupiter flyby formally ended on January 2, 1974. *Pioneer 10* crossed Neptune's orbit in May 1983, and for many years was the most remote man-made object in the Solar System. The last signal from the spacecraft was received on January 23, 2003. The *Pioneer 10* is heading for the star Aldebaran in the constellation of Taurus. It will take the spacecraft over two million years to reach it.

Saturn — *Pioneer 11*, *Voyagers 1* and *2*, and *Cassini* — studying this large planet, including its rings and moons. In 1986, *Voyager 2* flew past Uranus and detected ten new moons. Three years later, *Voyager 2* flew past Neptune, giving the first good view of its clouds, rings, and moons. In January 2006, NASA's *New Horizons* probe blasted off from Cape Canaveral on its nine-year journey to distant Pluto. The probe will speed past Pluto on July 14, 2015, and then on through the Kuiper Belt over the following years.

BENEFITS FROM SPACE

Although it may not be immediately apparent, NASA space launches from the Kennedy Space Center that put spacecraft into Earth's orbit, astronauts on the Moon, and probes on far away planets benefit mankind. Advances in technology that were needed to send humans into space, keep them alive, and return them safely to Earth are being used today in all sectors of our daily lives. The more we learn about the other planets, the more we understand about Earth.

New technologies are developed by NASA, born out of necessity to meet the unique challenges of space exploration. NASA has been issued over 6,300 patents, nearly one in every thousand patents issued by the U.S. Patent Office since 1790. Civilian use of space technology is limited only by the imagination and needs of the general public.

All professions have in some way been affected by new advances in materials, hardware, and procedures developed by NASA. Spin-off technologies adapted for civilian uses have found their way into new life-saving medical instruments, stronger and lighter construction material, food packaging, education, transportation, and astronomy. Thousands of computers are used in NASA's systems, and many advances in computer technology are direct spin-offs of NASA's applications.

Other examples of our space program's influence on our everyday life include:

✦ NASA-made polymers are widely used in a variety of commercial applications.

✦ The scope, clarity, and reliability of our long-distance telephone system emerged from thirty years of NASA development of satellite technology.

✦ Weather satellites orbiting the Earth transmit information that makes long-range weather predictions possible and local TV weather reports more accurate.

✦ NASA research into advanced ceramic led to teeth-straightening braces made of nearly invisible translucent material and fire-resistant suits that protect firefighters and racecar drivers.

✦ Earth surveys taken by the *Skylab Space Station* uncovered new deposits of oil, ore, and water.

✦ Following NASA's lead in developing new technologies for computer software simulations for integrating payloads and vehicles, clothing manufacturers utilized this technology to enhance the competitiveness of the U.S. apparel industry.

✦ A swimsuit designed by NASA and Speedo, which reduces drag by twenty-four percent, was widely used at the 2008 Summer Olympics.

✦ Precision optics also helps in the manufacture of computer chips.

✦ To keep sensitive space electronics at the correct temperature while awaiting launch, NASA, with the help of the Florida Solar Energy Center, developed a new heat pipe technology. A Georgia candy cane company adapted that heat pipe technology and hardware to store their candy cane products in a year-round cool place.

✦ An infrared technology developed to detect the birth of Stars is now used as an ear thermometer.

✦ NASA's success with the Space Shuttle program has led to the birth of a private space industry. Several entrepreneurs are investing time and substantial personal fortunes in building a new generation of rockets and vehicles for commercial space travel.

✦ The technique of cutting thin grooves across concrete to improve tire traction in wet weather is now used on highways and overpasses nationwide.

✦ Leftover rocket fuel is fashioned into a flare that destroys buried land mines.

✦ Heat protection blankets designed to keep the Space Shuttle cool during re-entry into the Earth's atmosphere were approved for use in NASCAR racing cars to help keep the driver cool. NASCAR driver Rusty Wallace demonstrated the blanket's effectiveness with controlled tests at the Daytona International Speedway.

✦ Methods to grow plants in space have produced world-record crops on Earth using hydroponics.

✦ In the absence of gravity new allows can be made. The new and improved materials that can be made in space may very likely be the foundation of important future industries on Earth.

✦ A NASA computer program to see how rockets and spacecraft would stand up to stress is now used for airplanes, cars, bridges, and skyscrapers.

✦ New industries have started as a result of advances in cryogenics, the study of ultra-low temperatures, required to deal with liquefied gases used as rocket fuel. Among the beneficiaries are hospitals, which require large amounts of liquid oxygen, nitrogen, and helium.

+ Insulating material developed for satellites is now used in homes and oil company pipelines.

+ Special photographic techniques used in Moon and planet research are now used in the search for oil and minerals.

+ Sun-powered homes are designed using technology that helped power satellites with solar energy.

+ Satellite communications technology has been applied to producing bright wall coverings that have additional insulation and lasts longer.

+ NASA and the Department of Energy and the Interior have been developing a special satellite that would collect solar power and beam it to Earth. In the future, cities may be powered by such satellites.

+ Electronic engine diagnostic equipment, widely used by auto mechanics, was adapted from the complex systems that automatically made sure everything in the NASA rocket/spacecraft was "GO."

+ Nitinol, a wire material developed for NASA, has been adapted by a firm that produces orthodontic equipment to make more efficient braces for teeth.

+ NASA's technology that monitored the health of astronauts in Space is now being used by doctors to detect the presence of dangerous bacteria in the body and in devices for testing hypertension.

+ Emergency vehicles use remote patient monitoring methods that were developed by NASA for monitoring astronauts in Space.

+ Space technology has been used to find ways of cutting down the danger of skidding on highways and airport runways.

+ To improve airline safety, many pilots are now trained with easy-to-use 3-D simulation software based on NASA space-related simulators.

+ Space technology helped perfect the nickel-zinc battery used in many vehicles.

+ The catalytic converter in many automobiles may have been derived from an ultra-compact catalytic converter developed for spacecraft life support.

+ Thanks to technologies developed by NASA, some pacemakers now have rechargeable, long-life batteries. They use a single chip derived from micro-miniaturization technology developed to fit complex equipment on small spacecraft. The same system that allows NASA engineers to reprogram satellites allows a physician to reprogram a pacemaker inside the patient's body without surgery.

+ Wind tunnels at NASA's Langley Research Center helped Cessna develop an efficient high-speed, business-class jet that races across the sky at an altitude of 51,000 feet at a speed of 0.92 — nearly the speed of sound.

+ The Chrysler Corporation, which worked on several NASA test launch vehicles, learned a lot about computer analysis of engine emissions. "Lean burn" engines are already being used in the production of current automobiles.

+ NASA-developed lightning detectors are now being adapted for commercial aircraft.

+ Oil seals developed to withstand the extreme pressure of Space flight are now used in automobiles.

+ A paint company borrowed from NASA's work with "inorganic paints" the method of producing a heat-resistant paint.

+ All the advances in aerodynamics, jet propulsion, fuels, communications, and other aspects of aviation flight can be traced directly to the Space program.

+ The first balloonist to cross the Atlantic Ocean said they would not have attempted the trip without a position-locating device developed by NASA.

+ Flotation devices used to support Space capsules following splashdown have been adapted to make weatherproof lifesaving craft that cannot capsize or sink.

+ Space research into early warning systems improved the state of the art, leading to the development of inexpensive, smoke detector devices used in homes and businesses.

+ A spin-off of research on helicopter rotors for NASA (and the military) led to a new line of acoustic guitars.

+ To keep fit while they were encapsulated in Space for days at a time, *Apollo* astronauts pulled on ropes from an aluminum cylinder hooked to the wall of the spacecraft. Similar devices are widely used by athletes.

+ Several brands of tennis rackets have been produced from graphite and other "composite alloys" developed to strengthen and lighten the spacecraft.

+ A NASA-commissioned design that helped astronauts breathe on the Moon when they left the Lunar Module has been converted into a lightweight oxygen backpack for firefighters.

+ Monitoring devices left on the Moon have resulted in burglar alarms for use on Earth.

+ Efforts to protect and make things more comfortable for the astronauts have led to a number of advances in furniture and sports equipment.

- A microscopically thin film that can be sprayed on phonograph records lubricates the surface to protect the grooves against the harder stylus. It is a spin-off of the technology developed for the moving parts of satellites.

- NASA spin-off developments include thermal insulation used in toxic chemical protection, survival gear, clothing, and fire protection equipment.

- Robotic manipulators with industrial applications were used to produce devices to aid disabled and paraplegic people in everyday tasks.

- Toxic chemical strippers were patterned after water-jet coating strippers designed to clean the Space Shuttle's reusable rocket boosters.

- Radiation blockers for sunglasses screen out harmful waves to enhance eye protection and reduce eye damage.

- Ergonomic seats designed for astronauts have resulted in desk chairs that reduce work-stress injuries, support chairs for paraplegics, and chairs for the elderly.

- Satellite technology has helped produce pump therapy to control injection of drugs for diabetes patients.

- Improved prosthetic devices were adapted from a design meant to help keep the Space Shuttle's fuel tanks super cold.

- The *Skylab Space Station* temperature control systems led to computerized water heating systems.

- A device developed for spacecraft location is now attached to large fishing nets and emits a warning signal audible to dolphins, who can then avoid being caught in the nets.

- Satellite remote sensing devices led to the development of Magnetic Resonance Imaging (MRI) systems.

- Satellite map resources help provide vital data about water and food conditions.

- Exercise machines developed for astronauts on Space Stations are now used by health-conscious humans throughout the world.

- Space technology helped introduce infrared mapping for urban planning.

- Spacecraft electrolytic water filters led to the development of domestic water purifiers.

- Satellite imagery has been used to detect major archeological finds, for example, the lost city of Ubar in the Sahara desert in Africa.

- Space technology has helped produce fire scanners that detect sources of visible and invisible flame.

- Plants that purify sewage are cultivated for use as purifiers in sewage lagoons.

- Satellite programs for atmosphere and climate monitoring led to search-and-rescue technology, which has saved many lives.

- Running shoes have been adopted from spacesuit technology.

- NASA developed weather-resistant coating is widely used in construction projects, including protection for the surface of the Statue of Liberty.

- Laser eye surgery was first used during satellite atmospheric studies.

- Technology developed for use in the Space program has led to improved design efficiencies, which result in creating products that conserve energy and pollute less.

- A material for mattresses, called Tempur, that relieves back pain was derived from technology to reduce gravity forces experienced during rocket/spacecraft launches.

- A NASA developed anti-fogging spray is now used on camera lenses, pilot and motorcycle helmets, and windows.

There are also benefits from space technologies that are not related to direct spin-off products or procedures. Only in the eyes of the workers at NASA facilities throughout America will you see the mixture of pride and humility that comes from watching a NASA spacecraft go into Earth orbit, a man walking on the Moon, a safe recovery of an astronaut returning from space, a space probe transmitting new unseen before photographs of a far away planet, or a successful Space Shuttle leaving the launch pad. These NASA workers realize that these events will eventually affect the lives of everyone on planet Earth.

It would be impossible to count how many students stay in school to study hard subjects like mathematics, science, and physics because they dream of working in space technology someday. We probably all know of young boys and girls who want to grow up and be astronauts.

THE FUTURE

There are several reasons why countries launch humans into space atop pillars of fire. Since the dawn of the Space Age and the Cold War rivalry between the United States and the former Soviet Union, scientific discoveries, international prestige, geopolitics, national security, and economic benefits all have been invoked to build public and political support for sending humans into orbit.

Today there is an extremely high cost associated with sending probes to far away planets, orbiting satellites around the Earth, and putting robot rovers on planets. When humans instead of robots are added to the passenger list, the expenditures rise exponentially. Scientists need to make sure they leave, arrive, and return safely. When talking settlements on the Moon or Mars, the expense must be added of generating power, shipping gear, and designing spacesuits and tools that can withstand unfathomable harsh environments. These items don't just come off the shelf at your local hardware store. The estimates for a single lunar outpost vary from tens of billion to hundreds of billion dollars.

Many countries are considering collaborating on future space missions. The Soviet Union and the United States started cooperative space ventures in 1975 with the Apollo-Soyuz Test Project and then again in 1995 with the Mir-Space Shuttle Cooperative Project. To a certain extent, of course, we already have an active partnership project. The *International Space Station*, by name and definition, is a collaborative project between U.S., Russia, Canada, Japan, and several European countries. China is a new entrant in human spaceflight. Over the past six years, China has launched three orbital missions. The latest venture, in December 2008, carried three taikonauts (astronauts) into orbit. Two of them conducted China's first spacewalk. On October 22, 2008, India sent a probe to the Moon and plans to launch its first astronaut sometime before 2015. If successful, it would become the fourth country with the ability to launch humans into orbit.

The Cassini-Huygens mission to Saturn is another cooperative venture between NASA and the European and Italian space agencies. The United States also had a key instrument on the probe India recently ferried to the Moon.

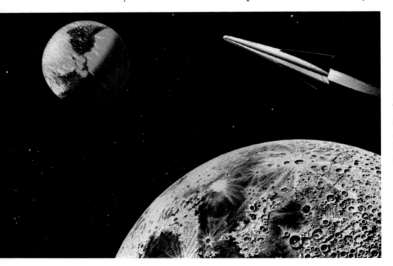

Future Space Travel. This new view shows an artist's conception of a future spacecraft leaving the Moon for a return trip to Earth.

Researchers continue to look for water on the Moon. NASA's *Deep Impact* and *Cassini* spacecraft missions, along with India's lunar probe *Chandrayaan-1*, detected a small amount of water on the Moon. In late 2009, NASA sent the *Lunar Reconnaissance Orbiter* (LRO) and the *Lunar Crater Observation and Sensing Satellite* (LCROSS) spacecraft on water-sensing missions to the Moon.

Future Russian space plans include *Phobos-Grunt*, a return probe to gather rock and soil samples from the Martian moon Phobos; *Luna-Glob*, a lunar probe that is scheduled to go up in 2012; and a *Venera-D* probe to map Venus, slated for 2016.

However, astronauts may soon not be the only people orbiting Earth. On December 7, 2009, billionaire Richard Branson's Virgin Galactic Company announced that commercial space travel will be available in the future from the company's launch complex in southern New Mexico. Anyone with $200,000 will be able to make a two- and a half-hour suborbital trip in the Virgin Spaceship Enterprise, a six-passenger, two-pilot spaceship. Three hundred people have already booked a flight on one of Virgin's five spaceships.

Current Administration's Plans

In 2010, President Obama announced that he wanted America's future exploration to be more global, with multiple countries teaming up to send astronauts to an asteroid by 2025, orbit Mars by the mid-2030s, followed by a landing on Mars. Under Obama's plan, NASA would get out of the rocket-making business and would instead contract out rocket design, test, and fabrication jobs to private industry. However, NASA would still remain in the pilot's seat for oversight, safety, mission assurance, and launch operations. NASA would partner with industry to spur innovation with prudent, but minimal intrusion. Current plans include a new heavy-lift rocket that would take future astronauts deep into Space. These heavy-lift boosters could send astronauts to places such as the Moon, near-Earth asteroids, the moons of Mars, and eventually Mars itself. Advances in rocketry could open up space travel

NASA's Space Shuttle program will come to an end in 2011. Until a commercial rocket is ready to haul passengers, American astronauts will continue to hitch rides to and from the orbiting International Space Station on the Russian Soyuz spacecraft. Each seat costs NASA tens of millions of dollars.

Under President Obama's new plan for America's space program, future planetary science research and robotic exploration could benefit. Healthy levels of funding could be available for new robotic missions to the Moon, asteroids, Mars, and Jupiter. Expanded international space partnerships would be developed, with NASA taking a leadership position.

CAPE CANAVERAL: BEFORE THE ROCKETS

IT IS GENERALLY BELIEVED THAT THE CAPE CANAVERAL AREA, AS IT IS KNOWN today, played host to a large variety of flora and fauna, including just about any creature that could fly, walk, or crawl to its shores. The ancient Indians that populated all of Florida were the Paleo Indians, who, using stone-tipped spears, hunted the great wooly mammoths, mastodons, saber-toothed tigers, and other large animals. Perhaps some of these animals ambled across the coastal shallows to graze on the ancient Cape Canaveral landscape.

Indians in the Archaic Period that followed were less dependent on large animals. They killed smaller animals with darts and spears and gathered produce from the land and sea.

The Indians found by the Spaniards in Florida were a wild and savage people. Archaeological discoveries have indicated that a wide range of Native American groups left their mark in the Cape Canaveral area. Literally hundreds of middens are located in Cape Canaveral and the surrounding areas. Each midden is typically several feet high and contains the remains of coquina, whelk, oyster, and clamshells. Some of the middens contained bones and artifacts.

It is generally believed that two main groups of Native Americans populated the Cape Canaveral area leading up to Colonial times. These are the Ais and Timucua Indians, both of whom frequented the area due to its local abundance of seafood and edible vegetation.

The Ais Indians are thought to have populated the coastal area along the Indian River south to the St. Lucie River and extending perhaps as many as thirty miles inland. A branch of the tribe has been cited as the rescuers of Quaker Jonathan Dickinson when his ship wrecked off Hobe Sound in East Florida. Despite the rescue, Dickinson later recounted in his book, *God's Protecting Providence*, that the Indians stripped Dickinson's party of men, women, and children of their clothing. After months of wilderness hardship, they made their way to St. Augustine. The Ais Indians were fiercely warlike and known to be cannibals. The Ais hated the Spanish and were the chief reason the Cape Canaveral area was not colonized by Spanish settlers. A large number of Spanish shipwrecks were plundered by the Ais. They also salvaged tons of Spanish gold and silver, which is periodically discovered hidden in the middens.

The Timucua Indians populated a large area extending from Cape Canaveral north to Georgia. They were docile in comparison to the Ais, and were primarily hunters and fishermen. They also raised crops. By the mid-1700s, the Ais and Timucua Indian tribes, as well as other tribes in Florida, were almost all gone as a result of settlement battles, disease, and forced labor (slavery). By the time Spain turned control of Florida over to the British in 1763, few Indians remained.

FAR LEFT: Paleo Indians. In Ancient times, mammoths, mastodons, bison, and saber-toothed cats migrated through Florida and were hunted by Paleo Indians. This postcard, published by the Brevard Museum of History and Natural Science in Cocoa, shows a colorful diorama of Cape Canaveral's first inhabitants attacking a Florida mammoth. Visitors can handle fossils, specimens, and artifacts from the animals and cultures that lived in the Cape Canaveral area over thousands of years.

NEAR LEFT: Ais Indians. The Ais Indians were a distinct tribe from their northern Timucua Indian neighbors, and primarily inhabited the Indian River and Cape Canaveral areas.

CAPE CANAVERAL BECOMES A U.S. POSSESSION

Cape Canaveral, along with the rest of Florida, became a possession of the United States on July 10, 1821. A year later the United States made Florida a Territory and then began to prepare Florida for statehood. In 1845 Florida became the twenty-seventh state of the United States.

In the early 1800s, 23-year-old Douglas D. Dummit established the first permanent settlement in the Cape Canaveral area. By 1828, Dummit shipped commercial quantities of oranges northward along the Indian River. Much of the original Dummit Grove is today located on the Kennedy Space Center property.

In the 1840s, the first group of settlers established permanent residence on geographic Cape Canaveral. These settlers occupied just a few households, but were able to maintain a self-reliant existence that at the time was a hostile environment marked by brutal heat, plagues of mosquitoes, a challenging sand and scrub environment for growing crops, and most of all isolation from other people.

CAPE CANAVERAL LIGHTHOUSE

The U.S. government selected Cape Canaveral as the site for a permanent lighthouse. Work on a sixty-five-foot brick tower near the shore began in 1843, although it was January 27, 1848 before the station was fully established. Before long it was realized that a taller tower and stronger light was needed. By 1873, a 168-foot cast-iron lighthouse was completed. Its sloping sides, lined inside with brick, were painted with alternating broad white and black bands.

Within a decade of its completion, erosion threatened the Cape Canaveral light. In 1893, workers unbolted the massive plates and reassembled them more than a mile inland. In the meantime, the old brick tower was dynamited

and the resulting fragments were used to make a concrete foundation for the relocated lighthouse. The beacon went back into service in July 1894, flashing a powerful white light through a first-order Fresnel lens.

The Cape Canaveral light was automated and unmanned in 1967. The light remains active, although the huge French optic was removed in 1993. It is displayed on the grounds of the Ponce de Leon Light, near Daytona Beach.

GROWTH AROUND THE CAPE

Mainland areas to the west of Cape Canaveral experienced a steady growth through the 1890s, when Henry Flagler's Florida East Coast Railway was extended into the area. By June 1893 the railroad reached the city of Titusville, which was a popular port and featured a mule-driven railroad that carried goods to western settlers in what is now the Orlando area. The Flagler railroad was quickly extended southward along the western bank of the Indian River through the cities of Cocoa, Rockledge, Eau Gallie, and Melbourne. The Titusville, Cocoa, and Melbourne areas soon emerged as major centers of local population.

Cape Canaveral Lighthouse. The Cape Canaveral Light, located on the tip of the Cape, is maintained by the U.S. Coast Guard as an official navigation reference for aircraft and ships. Natives occasionally try to convince gullible, gawking tourists that the remote Cape Canaveral Light is the next missile awaiting launch.

Tourists Start Visiting Cape Canaveral. After Cape Canaveral became America's guided missile shooting gallery in the 1950s and space launching center in the 1960s, the growth of the area was rapid. Local businesses started advertising the Space Coast as a popular place to visit. Shown is a postcard from the popular Bernard's Surf Restaurant in Cocoa Beach. In 1962 it was the author's first time to see Bear Steak on a restaurant menu. The astronauts and space center workers were often seen visiting this restaurant.

Ramon's of Cocoa Beach. Throughout the 1960s, Ramon's restaurant in Cocoa Beach was very popular with the astronauts, U.S. Air Force, NASA, and other personnel working on the Mercury, Gemini and Apollo programs. Ramon's, who advertised itself as "The Toast of the Coast," had three dining rooms and two cocktail lounges. Throughout the restaurant was space related memorabilia.

However, Cape Canaveral remained isolated, accessible only by boat. In 1923, the first bridge was built from the mainland eastward to the Atlantic Coast. The eastern terminus of this bridge was incorporated as the city of Cocoa Beach in 1925. After Cape Canaveral became America's missile test range and America's launching pad for space vehicles, the population of the area increased from around 23,000 in 1950 to an estimated 250,000 in 1969.

BANANA RIVER NAS

Events that shaped the future of Cape Canaveral began just prior to World War II. In 1940 the Banana River Naval Air Station opened on a narrow strip of barrier island located roughly between the Melbourne and Cocoa areas. It covered 1,791 acres and was roughly four miles long and more than a mile wide. During World War II this naval air station supported coastal seaplane patrol operations and operated a PBM seaplane pilot training program and advanced navigation school. The station was officially deactivated August 1, 1947.

MISSILE TEST RANGE

In 1949, the closed Banana River Naval Air Station and land on Cape Canaveral became an Air Force missile test range. "The Cape," as most missile engineers and people working with missile-launching activities call it, became a 5,000-mile shooting gallery of open water for America's then new long-range guided missiles. In the mid 1950s, the Army Corp of Engineers and Duval Engineering Company of Jacksonville began work on constructing the first permanent access road, instrumentation sites, military offices, and launch sites in Cape Canaveral. The first area developed for launch operations became known as Launch Pads 1, 2, 3, and 4. These launch pads were built near the Cape Canaveral Lighthouse. The first tracking stations were also constructed on the Bahamas Islands. The range tracking stations eventually stretched across the Atlantic Ocean and South Africa and into the Indian Ocean.

Space Launch Watching Sites. This view is from Space View Park on the Indian River in downtown Titusville, directly west of the Kennedy Space Center's Vehicle Assembly Building (VAB) and Launch Pad 39A. When this view was taken, the Space Shuttle *Atlantis* was sitting on the launch site. The VAB is seen in the center of this view. On launch day, hundreds of people were watching the launch from this site.

Canaveral Pier. Originally built in 1962, the Canaveral Pier (also called the Cocoa Beach Pier), which stretches eight hundred feet over the Atlantic Ocean, was a popular fishing spot. Now a gathering spot of restaurants and shops, the pier is the social hub of the college crowd during Spring Break and a watching spot for missile launches.

Project Mercury Exhibit. The Project Mercury Exhibit in Space View Park in Titusville has plaques of the Mercury astronauts, along with their handprints.

Cocoa Beach. The shoreline of the Atlantic Ocean is not only a popular vacation spot, but also provides an excellent view of launches from Cape Canaveral. This view shows how Cape Canaveral (top) juts out from the Florida coastline.

PORT CANAVERAL

The deep-water port of Port Canaveral was constructed to allow the berthing of range instrumentation and cargo ships; however, it was later expanded to service ballistic missile submarines and commercial vessels. The protected harbor of Port Canaveral, located at the southern end of Cape Canaveral, is within sight of the launch pads at The Cape. Today, Port Canaveral is one of the largest cruise ship ports in the world and docks some amazing cruise ships. The Carnival Cruise Line, Disney Cruise Line, and Royal Caribbean International have mega ships that offer weekly cruise trips to the Bahamas and Caribbean Islands. Over the years, several partial-day gaming ships have also operated out of Port Canaveral. The Sterling Casino Lines gaming ship, the largest gaming ship in the world, once sailed on twice-daily gambling excursions out of Port Canaveral. Jetty Park, located on the eastern side of the Port, is an excellent location to watch the glamorous vessels arriving and leaving the Port, as well as missile launches from Cape Canaveral.

Port Canaveral, Florida. Today, Cape Canaveral is better known as a spaceport than a seaport, but Port Canaveral is increasingly busy as a staging area for cruise ships. Visitors fascinated with space travel can begin or end their cruise with a visit to the Kennedy Space Center and Cape Canaveral Air Station.

Gaming Ship Arriving at Port Canaveral. Sterling Casino Lines began sailing the *Ambassador I* gaming ship out of Port Canaveral on September 8, 1998. It was replaced by the five-deck, 11,940-ton *Ambassador II* (shown), the largest casino ship in Florida. It had fifty gaming tables and 1,000 slot machines and was one of Florida's most luxurious casino ships. This ship, which carried 1,800 passengers, stopped sailing in 2008.

Caribbean Bound. Hundreds of passengers on this Disney Cruise Line cruise ship have a good view of their Caribbean-bound vessel leaving Port Canaveral.

Leaving Port Canaveral. The people fishing on the Jetty Park jetty have a good view of the Carnival Cruise Line *Glory* leaving Port Canaveral for the Caribbean. This location also provides a view of Cape Canaveral's missile launches.

THUNDER AT THE CAPE

The First Rocket Launch

THE FIRST MISSILE FLOWN FROM CAPE CANAVERAL WAS A TWO-STAGE BUMPER 8, consisting of a V-2 rocket as the first stage and an Army WAC-Corporal rocket as the second stage; the date was July 24, 1950. The site for Bumper launches was Launch Pad 3, which had been hastily, but thoroughly, completed to support the launches. Five days after the first Bumper 8 launch, the Bumper 7 missile was launched. These launches successfully demonstrated the feasibility of near-horizontal liquid fuel rocket staging.

Maximum range of these Bumper flights was about 150 miles. Although seemingly insignificant by modern standards, the stage was thus set for many milestones to come, with the range of Cape-launched rockets ultimately extended to send men to the Moon and spacecraft to the outer reaches of the Solar System. In the years since, the Air Force, Army, Navy, and NASA have launched thousands of missile- and satellite-carrying launch vehicles.

Tracking Missiles

Missile tracking stations are located on islands in the North and South Atlantic Oceans, augmented by a fleet of ocean-going tracking ships and specially instrumented aircraft. Tracking a missile as it flies downrange provides needed performance data and safety information. If a missile should fail to obey commands and begins to wander from the planned path, the Range Control Officer can send corrective signals to direct the vehicle's guidance system to bring it back or send a signal to an explosive charge on the missile, therefore destroying the missile while it is still over water.

Patrick Air Force Base

A few days after the historic Bumper missile launches from Cape Canaveral, the missile launch facility received the name it holds today. It was renamed Patrick Air Force Base in honor of Major General Mason M. Patrick on August 1, 1950. The Cape Canaveral launch area became simply an extension of Patrick Air Force Base, which was declared a permanent military installation on December 24, 1952.

Expanding Operations

Cape Canaveral began to grow rapidly. As new ballistic missiles were introduced and tested, new facilities at The Cape were constructed out of necessity. There once were some twenty active launch pads on Cape Canaveral and it was not at all uncommon to witness several launches in the space of one week.

First Successful Rocket Launching from Cape Canaveral. At 9:29 a.m. on July 24, 1950, a makeshift rocket on a crude launch pad blasted off from Cape Canaveral. Control centers were located in trenches, a tent, and an old Army tank. From the randy road leading to the launch site, snakes and alligators watched with the thirty-man launch team as the Bumper missile zoomed ten miles into the air. A few minutes later it dropped fifty miles away into the Atlantic Ocean. On July 29, 1950, five days after the initial launch, a second Bumper missile was launched at the Cape. The Bumper was composed of a German V-2 rocket similar to those that bombed London in World War II, and topped with a smaller rocket, called the WAC Corporal.

Patrick Air Force Base. The Air Force Missile Test Center's Technical Laboratory at Patrick Air Force Base, located along the Atlantic Ocean just south of Cape Canaveral and Cocoa Beach. Highly trained personnel working in this building evaluate data from all missile and satellite launching at the Cape. In front of the building, which is located on Florida Highway A1A, are several missiles developed by the U.S. Air Force.

Aerial Views of Gantry Row. At one time there were some twenty active launch pads on Cape Canaveral. This 1964 view (looking north) shows several of the launch pads. This is where the launches for the Mercury and Gemini projects took place. The Atlantic Ocean is shown on the right. In the distance is the Kennedy Space Center with the Vehicle Assembly Building (VAB) and Launch Pads 39A and 39B.

WINGED MISSILE LAUNCHES

The earliest research-and-development (R&D) vehicles flown on the Atlantic range were winged missiles, the Martin Company's 650-mile-range Matador and the 5,000-mile Snark, both of which basically were pilot-less bombers.

The Matador was the Air Force's first missile program to be flight-tested and the first program to be completed. Between June 1951 and November 1956, 154 Matadors were flown on R&D flights from The Cape.

Snark, the first long-range U.S. missile, made ninety-seven flights down the Range between August 1952 and December 1960.

Other Air Force pilot-less winged missile systems included the Navaho, a two-stage vehicle based on a German World War II concept of a skip-glide bomber that made twenty-six flights during 1955-1959; the Martin Company's Mace, an improved and longer-range version of the Matador that made forty-four R&D flights during 1959-1963; and the pilot-less interceptor Bomarc, which made seventy flights.

On June 3, 1954, Cape Canaveral supported the first attempted recovery of a winged missile that flew a programmed pattern and then returned to The Cape for refurbishing and reuse. A Snark missile was successfully guided for landing on the Cape Canaveral Skid Strip, but the missile's skids failed and the vehicle crashed and exploded upon contact. The Snark missile employed metal skid plates, but no traditional landing gear. A large number of Snark missiles were successfully recovered this way. In 1956, a Navaho missile was successfully launched from and recovered on the skid strip. Later Navaho missiles employed landing gears that allowed them to land similar to traditional aircraft. In 1963 the Cape Canaveral Skid Strip was also used as an auxiliary landing strip for Patrick Air Force Base.

Martin "Matador", Pilotless Bomber, being launched from launching site Cape Canaveral, Florida

Matador Launch. The First Matador pilot-less bomber was launched June 20, 1951. Ten years later the U. S. Air Force crew launched the last Matador. Over 150 Matadors were flown on Research & Development (R&D) flights from Cape Canaveral. In addition, U.S. Air Force crews who later stood guard in Europe with two squadrons of the weapon flew over 130 Matadors as training vehicles.

Snark Launch from Cape Canaveral. The first Snark was launched August 29, 1952. The Snark, similar to the Matador missile, had a range of 5,000 miles.

Mace Missile Launch from Cape Canaveral. An Air Force TM-76B MACE tactical range ballistic missile built at the Baltimore Division of The Martin Company blasts out of a prototype hard site at the Air Force Missile Test Center at Cape Canaveral on October 29, 1959. The Mace missile was an improved and longer-range version of the Matador.

Final Flight of Navaho from Cape Canaveral. This Navaho missile was launched November 18, 1958, by North American on Pad 9. It never became an operational weapon.

Launching of the First Bomarc Missile. On September 10, 1952 the first USAF Bomarc missile was launched at Cape Canaveral. The Bomarc, shown here blasting off, is a long-range area defense guided missile designed to destroy enemy aircraft and missile before they approach U.S. target areas.

EARLY BALLISTIC MISSILES

Ballistic missiles, which fly like an artillery shell in a long arc from launch stand to target in minutes, replaced the older winged cruise missiles that flew like conventional aircraft.

Redstone

The Army's 200-mile range Redstone missile, built by Chrysler Corporation, signaled the end of the winged missiles when it first flew on August 20, 1953. It made eighteen R&D flights, was used as a booster for other programs in nine launches, and served the Project Mercury manned space flight program six times. Two of the Redstone vehicles lifted astronauts Alan B. Shepard and Virgil I. Grissom briefly into space.

Jupiter Missile. The U.S. Army developed this Jupiter missile shown in the service tower at Cape Canaveral. It's being checked out before a test flight at the Atlantic Missile Range.

Juno II. The Air Force's Juno II rocket is prepared to launch by missile workers at the Cape Canaveral launching site.

Redstone Being Fueled. The first Army Redstone missile was launched August 20, 1953. Shown is the Redstone, which was test-flown over the Air Force Missile Test Center's Range, being fueled at the Cape Canaveral launching site. The gantry service tower standing beside the missile was used in pre-flight preparations.

Jupiter and Juno

The U.S. Army's Jupiter, built by Chrysler Corporation, was assigned to Air Force units for operational deployment. The 58-foot-tall, 110,000-pound missile made sixty-five R&D crew training and space test launches from 1957-1961. The Jupiter missile also gave two monkeys, Able and Baker, a brief ride into space on May 28, 1959.

The Jupiter C, a modified Redstone, was used to orbit *Explorer 1*, the first U.S. satellite, on January 31, 1958. This vehicle also went on to launch *Explorer 3* and *4*.

Juno II is a Jupiter missile with a modified upper stage. This vehicle was launched ten times between 1958 and 1961, and placed four satellites in orbit: *Pioneer 4*, which went into orbit around the Sun in 1959, and *Explorers 7, 8,* and *11*.

Army Redstone Missile. On September 17, 1958, this Redstone missile was being moved to its launch site.

Juno II Refueling Operation. The Juno II vehicle, similar to the missile that later launched space probes, is shown on its launch pad at the Cape Canaveral Atlantic Missile Range. This missile is being refueled with liquid oxygen for the launching.

A thrust augmented Delta rocket was launched from Cape Canaveral April 6, 1965, to orbit *Early Bird* of Communications Satellite Corporation, the world's first commercial satellite.

The Thor Missile. The Thor missile was an important member of America's missile arsenal. Here, the Thor is getting ready to be launched June 4, 1958. The Atlantic Ocean is seen in the background.

Thor-Able Ballistic Missile. The USAF's two-stage Thor-Able missile is readied for launch at the Air Force Missile Test Center's launching site at Cape Canaveral.

Thor-Delta Missile. The NASA Echo satellite, which unfolded from its canister into a 100-foot aluminized balloon to reflect radio signals, mated to the final stage of a Thor-Delta rocket at the Cape Canaveral launch site.

Lift-Off of Early Bird Satellite.

Thor and Delta

The Douglas Aircraft Company's Thor, a 65-foot-tall, 110,000-pound ballistic missile with a range of 1,700 miles, made its first flight in January 1957. It exploded in a tremendous fireball. The Thor went on to have over forty R&D launches at the Cape.

The Thor also became a launch vehicle carrier for satellites. The Air Force used the two-state Thor-Able to text experimental nose cone materials intended for Atlas and Titan missiles. The Air Force and NASA both used the Thor to launch lunar satellites and deep-space probes. The Thor-Able Star was used by the Navy for its Transit navigational satellite and the Air Force for the Army-Navy-NASA-Air Force (ANNA) geodetic mapping satellite.

NASA used the three-state Delta to launch various satellites, from eighty-pound *Explorers* to five hundred-pound Solar Observatories. The Thor was used as the first-stage in this composite vehicle.

Thor Leaving the Launch Pad. The first Thor long-range ballistic missile was launched at Cape Canaveral on January 25, 1957. Thor, a 65-foot-tall, 110,000-pound ballistic missile with a range of 1,700 miles, exploded in a tremendous fireball, However, many of the missile launches in the pioneering 1950s were less than perfect. It was not until the fifth vehicle that Thor scored a completely successful flight and then went on to score a creditable record during forty-nine launches at The Cape.

A Satellite Launch. A 385-pound geodetic Explorer spacecraft, designated GEOS-A, being launched from Launch Pad 17A atop a thrust augmented improved Delta missile on November 6, 1965. The spacecraft contains five geodetic instrumentation systems to provide simultaneous measurements that scientists require to establish a more precise model of the Earth's gravitational field, and to map a world coordinate system relating points on, or near, the surface to the common center of mass. This was the first launch for the improved Delta second stage.

Delta Vehicle Satellite Launch. An HEOS-1 satellite (Highly Eccentric Orbit Satellite) was launched aboard a 92-foot-high Delta rocket from Launch Complex 17B on December 5, 1968. Its purpose was to study cosmic radiation, solar winds and magnetic fields, two-thirds the distance from the Earth to the Moon. The 238-pound satellite was launched for the ten-nation European Space search Organization (ESRO).

Telstar I Being Launched. A Delta vehicle, launched from Cape Canaveral, for the American Telephone and Telegraph Company, to orbit Telstar I, AT&T's experimental communication satellite.

Telstar I Communication Satellite. AT&T's active communication satellite was orbited on a Delta rocket from Cape Canaveral on July 10, 1962, to provide an experimental telephone and television link between the United States and Europe.

Atlas

Five months after the first Thor and two weeks after the first Jupiter missile launches, the U.S. Air Force launched its first Inter-Continental Ballistic Missile (ICBM) on June 11, 1957. Atlas' first flight ended the same way the first Thor's did — in a deafening explosion. The second Atlas missile also exploded.

However, the Atlas, like Thor, went on to chalk up a fantastic record for its 85-vehicle flight test program that ended in 1962. Atlas scored a number of firsts: it was the first U.S. missile to reach full ICBM range of 5,000 miles (November 28, 1958); the first missile to fly more than 9,000 miles into the Indian Ocean (May 20, 1960, September 19, 1960, and July 6, 1961). Atlas flew itself into orbit on December 18, 1958, and on August 24, 1959, returned the first films ever taken of the Earth from an altitude of seven hundred miles.

Atlas Inter-Continental Ballistic Missile. The USAF Atlas Intercontinental Ballistic Missile has a speed in excess of 15,000 nautical miles per hour. It weighs 195,000 pounds when loaded for launching. At eighty feet long and a diameter of nine feet, it is shown in the beginning stage of being launched at the Cape Canaveral Missile Test Center.

Atlas ICBM Blasts Off from Cape Canaveral. The first Atlas was launched June 11, 1957. A similar type of vehicle launched the astronauts into their orbits around the earth.

Checkout of an Atlas ICBM. The Atlas Inter-Continental Ballistic Missile is shown on its stand at Cape Canaveral, launching site of the Air Force Missile Test Center. Technicians from Convair Astronautics, developers of the missile, perform checkout of the vehicle during the preflight stages, as they stand on the platform, as the red and white gantry crane is moved away.

Atlas-Agena Vehicle-Mariner I Satellite Launch. In 1962, the Atlas-Agena launch vehicle lifted into space, carrying the Mariner I satellite, on its way to the planet Venus.

The Atlas missile continued to be used as a booster for various missions. NASA and the Air Force used the two-stage Atlas-Able in a series of three unsuccessful Pioneer probes and lunar satellites in 1959 and 1960. NASA used the Mercury-Atlas vehicle to fly five unmanned Mercury spacecraft and then the *Friendship 7* capsule of astronaut John Glenn (February 20, 1962); the *Aurora 7* capsule of astronaut Scott Carpenter (May 24, 1962); the *Sigma 7* vehicle of astronaut Walter Schirra (October 3, 1962); and the *Faith 7* spacecraft of astronaut Gordon Cooper (May 15, 1963).

NASA used the two-stage Atlas-Agena to launch the *Ranger*, *Mariner*, the *Orbiting Astronomical Observatory* (OAO), the *Gemini* target vehicle, and *Lunar-Orbiter*. The Atlas-Centaur launched several *Surveyor* satellite probes to the Moon.

Research Flight Investigation of Reentry Environment. The Atlas missile lifted off Launch Pad 12 on May 22, 1965, carrying the second 200-pound Project FIRE spacecraft on a 5,000-mile ballistic trajectory to expand scientific knowledge of reentry heating. A velocity package using the solid propellant Antares II rocket motor added the speed needed to drive the reentry payload back into the atmosphere at 25,000 mph. The payload reached an apogee of approximately five hundred statue miles about 2,440 miles down range before starting reentry into the atmosphere. At the speed Project FIRE attained, the temperature of the gasses in the shock wave just ahead of the blunt reentry body approached 20,000 degrees Fahrenheit. As a result of data received from this flight, scientists were able to predict more accurately, for engineering purposes, the heating associated at speeds up to and beyond that of the Apollo command module on its return from lunar flights.

Launch of Orbiting Astronomical Observatory. NASA launched its most advanced unmanned spacecraft from The Cape on April 8, 1966, atop an Atlas-Agena launch vehicle. The Orbiting Astronomical Observatory was the first in a series of four designed to give astronomers their first sustained look into the Universe from above the obscuring and distorting effects of the Earth's atmosphere.

Atlas Centaur. The Atlas-Centaur is a two-stage launch vehicle, consisting of an Atlas and a Centaur second stage. The Centaur, powered by two 15,000-pound thrust (each) engines, burns liquid hydrogen, and liquid oxygen. The first launch from The Cape was May 8, 1962.

Titan I

The Martin Company's Titan I was a two-stage, ninety-foot tall, ten-foot diameter missile that weighed more than 220,000 pounds. This Air Force missile carried a nosecone nearly twice as heavy as that on the Atlas missile. The Titan I was first launched February 6, 1959, and completed forty-seven test flights. The Titan I was powered by two Aerojet-General Corporations first-stage engines of 150,000-pound thrust each, and a single second-stage engine of 80,000-pound thrust.

Titan II and Titan III

The 102-foot high, 300,000-pound, liquid fuel Titan II missile was first launched March 16, 1962. It was powered with Aerojet-General Corporation's engines that delivered a total thrust of 530,000 pounds. The National Aeronautics and Space Administration (NASA) used the Titan II to launch the manned *Gemini* capsules. The Titan II had only one failure in twenty-three launches. It once was the mightiest U.S. ICBM.

The U.S. Air Force developed the Titan III for various missions. The Titan III is a modified and more powerful Martin Company's Titan II, capped by a completely new upper stage, called the "transtage," which has two Aerojet engines of 8,000-pound thrust each.

Titan II Launch. The 98-foot-tall Titan II Inter-Continental Ballistic Missile (ICBM) lifts off on a flight down the Atlantic Missile Range. The Martin Company made the *Titan II*; its first launch at Cape Canaveral was March 16, 1962.

Titan I ICBM. The U.S. Air Force Titan I ICBM vents liquid oxygen minutes before launching at Cape Canaveral. The Titan I uses both liquid and solid propellants.

First Titan I Launch. This bird's-eye view of launching pads at Cape Canaveral shows the *Titan I* as it blasts off on a 6,000-mile trip down the Atlantic Missile Range. The Titan I, a two-stage ninety-foot tall, ten-foot diameter missile that weighed more than 220,000 pounds, was first launched at The Cape on February 6, 1959.

Titan IIIC. The *Titan IIIC*, at the time the world's most powerful missile, lifts off from the Cape on its inaugural June 18, 1965 launch. The *Titan IIIC* was followed by even more powerful versions: the *Titan IIIE-Centaur*, used to propel Voyager spacecraft toward the planets Jupiter, Saturn, Uranus, Neptune, and the farthest corners of the Solar System; and the *Titan 34D*, a more potent vehicle.

Pershing

The U.S. Army's two-stage, solid-propellant, 34-foot-tall, 10,000-pound Pershing ballistic missile was first launched at Cape Canaveral on February 25, 1960. Built by the Martin Company, over the next three years, fifty-two Pershing missiles were launched.

Pershing Missile. A tactical missile developed by the U.S. Army, the *Pershing* rises from its launch pad at Cape Canaveral. It is a two-stage solid propellant ballistic missile designed to replace the larger *Redstone*. The first *Pershing* was launched February 25, 1960. Three Pershing missiles were launched from Complex 16 on April 7, 1981. Launch times were 9:00, 9:19, and 9:40 a.m.

Minuteman

The U.S. Air Force's three-stage, solid-propellant Minuteman missile was designed to be stored in underground silos for long periods of time and be ready to launch in a second. The Minuteman replaced the Atlas, Titan I, and Titan II missiles. The almost sixty-foot-tall Minuteman weighs about 76,000 pounds and can place a small warhead on targets more than 5,500 miles away. The Minuteman's first test flight was February 1, 1961.

Minuteman Launch. A Minuteman missile is launched from the underground Silo at Cape Canaveral. The first Minuteman launch occurred February 1, 1961. The last launch of the three-stage, solid propellant Minuteman II ICBM occurred March 14, 1970 at the Cape.

Polaris

The U.S. Navy's Polaris Ballistic Missiles (later versions were the Poseidon and Trident) were designed to carry nuclear warheads that were to be launched from nuclear submarines. Several hundred Polaris R&D launches were made either from land pads at Cape Canaveral or from submarines lying submerged offshore.

U.S. Navy's Polaris Missile on Launch Pad 25A. The first Polaris missile was launched April 13, 1957. The *Polaris* was the U.S. Navy's version of the Army's Jupiter missile. The Poseidon, at thirty-four feet in length and 65,000-pounds in weight, was a newer, more powerful version of the Polaris ballistic missile. The Poseidon carried Multiple, Independently Reentry Vehicles (MIRVs), which were target-heavy with nuclear warheads. The three-stage Trident missile later replaced this missile.

THE SPACE AGE BEGINS

THE BIGGEST REVOLUTION IN THE HISTORY OF THE HUMAN RACE HAS TAKEN PLACE IN THE LAST SIXTY YEARS. We have been able to leave our planet and explore space. It has totally changed our lives — in fact, many of us would not recognize the world as it was before the launch of *Sputnik 1* in 1957. Now, flotillas of satellites circle Earth, beaming a cacophony of communications into our homes, while weather, resources, and even wars are surveyed from space. The attendant breakthrough in miniaturization and computer power can be appreciated by anyone with a personal computer, laptop, or a hand-held communications device.

Space is also a human frontier: hundreds of people have now flown in space, twelve have walked on the Moon, and sophisticated crafts have explored all the planets of our Solar System.

SATELLITES ORBIT EARTH

Wernher von Braun, developer of the German V-2 rocket during World War II, wrote a series of articles in *Collier's Magazine* in 1952 proposing and arguing in support of an American program of space exploration, first with instrumented satellites and then man to the Moon and Mars.

Three years later President Eisenhower announced that the United States would launch an artificial satellite in connection with the 1957 International Geophysical Year (IGY). The IGY was a cooperative scientific program of many nations wherein simultaneous geophysical and atmospheric measurements would be made around the world.

Sputnik 1 — The First Artificial Satellite

On October 4, 1957, a rocket thundered into space from the Soviet Union launch site near Tyuratam, northeast of the Aral Sea. Ten minutes later, the world's first artificial satellite, *Sputnik 1*, was in orbit. Its launch sparked off a major international rivalry known as the Space Race, as the United States fought to overcome this blow to its prestige while the Soviet Union strove to keep its propaganda lead.

Sputnik 1, from a Russian word meaning "satellite," was an aluminum sphere twenty-three inches in diameter and weighing 183 pounds. The very weight of the satellite was enough to scare the U.S., which was incapable of orbiting such large payloads, and it aroused fears of bombs raining from the skies on American targets.

Sputnik 1 orbited Earth every ninety-six minutes and emitted an eerie 'bleep-bleep' sound. Millions of people throughout the world heard its 'bleep-bleep' signals on radio and TV, and many satellite spotters saw the top stage of the carrier vehicle orbiting Earth like a brilliant star. *Sputnik 1* fell back to a fiery death in the atmosphere on January 4, 1958.

Sputnik 1 Stamps. These three stamps from Mongolia, Magyar (Hungary), and Romania depict the Soviet Union's *Sputnik 1*, the first man-made satellite to orbit the Earth, October 4, 1957. The 184-pound *Sputnik 1* satellite was a smooth aluminum sphere about the size of a basketball that contained a radio transmitter and batteries. Trailing behind it were four whip aerials of between 4.9 and 9.5 feet in length. As it circled the Earth every ninety-six minutes, its "beeping" signal delivered a message of apparent Soviet military superiority in space that caused major concern among Western nations. *Sputnik 1* transmitted for twenty-one days, but survived in orbit for ninety-two days before burning up during reentry on January 4, 1958.

Sputnik 2 Satellite. In 1975, Magyar (Hungary) issued this stamp illustrating the 1,118-pound Russian *Sputnik 2* satellite. *Sputnik 2*, launched November 3, 1957, carried a pressurized container with a dog, Geiger counters, and two ultraviolet and x-ray spectrophotometers. The scientific purpose of the mission was to gather data on the effects of weightlessness on an animal in preparation for a human flight. Though the satellite was not designed to return to Earth with its occupant, the *Sputnik 2* reentered the atmosphere April 14, 1958, after 2,570 orbits.

First Animal in Space. When the Soviet Union launched *Sputnik 2*, the attention of the world was fixed on the dog, Laika, that was aboard. She was the first living creature in space. Laika suffered no ill effects during launch, but died when the oxygen ran out in orbit. This stamp was issued by Mongolia in 1982.

Sputnik 2 — With A Passenger

A month later, on November 3, the Soviet Union launched its second satellite, the *Sputnik 2*, a 1,120-pound spacecraft carrying a dog named Laika — the first living creature to orbit Earth. Laika was kept alive in her pressurized capsule for a week while reports on her condition were automatically radioed back to doctors on Earth. Laika ran out of oxygen and died in space. The *Sputnik 2* burned up in the atmosphere on April 14, 1958. Laika, though, proved that living things could travel safely in space (as long as they have oxygen).

AMERICA'S FIRST SPACE SUCCESS

The United States was doing its first space work in the 1950s, but the work was moving slowly and the country was lagging far behind the Soviet Union. The main problem was that the three military services — the U.S. Army, U.S. Navy, and U.S. Air Force — were working separately and trying to beat each other to the honor of being first in space. Each service was developing its own rockets and was having trouble getting them off the ground, but then the Soviet Union's *Sputniks* went spinning around the Earth. The country found itself humiliated by being upstaged by the Soviet's leap into space. Top priority was given to space activities. While NASA was being formed, the three military services went on working separately.

Explorer 1

On January 31, 1958 the first U.S. satellite, *Explorer 1*, was launched by the Army at Cape Canaveral by von Braun and his team. The launch vehicle was a modified *Jupiter C/Juno 1* ballistic missile with three solid propellant upper stages. The thirty-pound *Explorer 1* orbited the Earth every 114 minutes.

It was a long, thin cylinder attached to the small top stage of its launch missile. The satellite carried scientific equipment that detected the now-familiar Van Allen radiation belts around Earth, named after the American physicist James Van Allen who devised the experiments on the satellite. The *Explorer 3* satellite launched a few months later and carried a tape recorder that proved the existence of these radiation belts.

Explorer 1 remained orbiting Earth (220-1,585 miles above the Earth) for more than twelve years, finally re-entering the atmosphere on March 31, 1970.

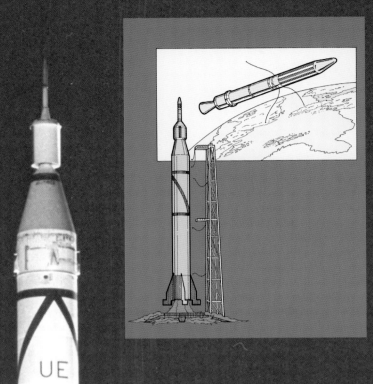

First U.S. Satellite to Orbit Earth. On January 31, 1958, the U.S. Army launched the *Explorer 1* satellite into orbit atop a Jupiter-C/Juno 1 rocket. *Explorer 1* was a cylindrical device 6.5 feet long and weighed 30.8 pounds. At its highest altitude, its orbit reached 1,585 miles. Its instruments confirmed the existence of the Van Allen belt, a layer of natural radiation that encircles the Earth at a height of six hundred miles. *Explorer 1* followed an elliptical orbit with a period of 114.8 minutes. Its final transmission came May 23, 1958. It reentered the Earth's atmosphere March 31, 1970, after completing more than 58,000 orbits.

Explorer 1 on its Way to Outer Space.
The *Explorer 1* satellite was launched on Launch Complex 26A at Cape Canaveral. It carried 17.6 pounds of instruments designed to collect data on cosmic rays, meteorites, and temperature.

MORE ADVANCES IN SPACE

The Soviet Union startled the world on May 15, 1958, by launching *Sputnik 3*, a 2,925-pound spacecraft carrying many geophysical experiments.

On December 18, 1958, the U.S. Air Force launched *Project Score*. An Atlas launch vehicle was placed in Earth orbit carrying a taped Christmas message from President Eisenhower. This was the first voice transmission from space.

By the end of 1959, the United States had launched eighteen spacecraft to the Soviet Union's six. Among them were further members of the *Explorer* series, which continued to study radiation around the Earth and even sent back the first crude pictures of Earth from space. The U.S. Air Force began experimenting with a series of recoverable spy satellites called *Discover*. Also around this time, the U.S was orbiting the first prototype satellites for weather, navigation, and communication purposes.

Soviet Probes to the Moon

The Soviet Union's lead in rocketry made itself felt again as attention turned towards the Moon. Three attempted American Moon shots had failed by the end of 1958. The Soviet Union lunar exploration program named *Luna*, consisting of a series of robot missions, was initiated. The missions were designed to study the characteristics of the Moon by orbital or landing vehicles carrying remote-controlled equipment. Between 1959 and 1976, at least twenty-four *Luna* missions were launched.

Luna 1 broke free from Earth's gravity on January 2, 1959, en route to the Moon. Programming errors with the launcher caused it to miss the Moon and go into orbit around the Sun like an artificial satellite. The *Luna 1* satellite still orbits the Sun every 443 days. Although the mission did not achieve all its planned objectives, it did discover the phenomenon of the solar winds and collected data about the Van Allen belt and the Earth's radiation belt.

On September 12, 1959 the *Luna 2* probe was launched and hit the Moon thirty-three hours later. For the first time, a man-made object had reached the surface of another body in space. Before *Luna 2* crashed, its instrument readings showed that the Moon had no detectable magnetic field. The probe had impacted as planned west of Mare Serenitatis, near the craters Aristides, Archimedes, and Autolycus.

On October 4, 1959, two years to the day after the launch of the *Sputnik 1* satellite, an even more important probe set off into space. This was *Luna 3*, whose mission was to photograph the far side of the

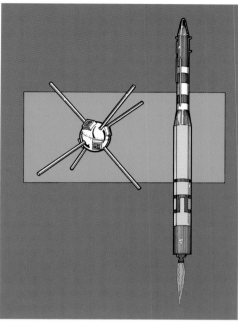

Launch of a Vanguard Missile. The first Vanguard missile was launched at Cape Canaveral December 8, 1956. This 75-foot-long vehicle was chosen by the Department of Defense to launch American satellites. The *Vanguard* flew a successful suborbital mission October 23, 1957. The first attempt to launch a satellite took place in December 1957, but ended with the vehicle exploding on the launch pad. After another failure in February 1958, the vehicle successfully launched the 4.5-pound *Vanguard 1*, America's second satellite, on March 17, 1958. Two more small satellites were launched before *Vanguard* was retired. Altogether, there were eight failures in twelve Vanguard launches.

Vanguard 1 Satellite. On March 17, 1958, the U.S. Navy successfully orbited America's second satellite, *Vanguard 1*. The satellite was a 4.5-pound sphere six inches in diameter. The launch booster consisted of a Viking rocket first stage, an Aerobee second stage, and the Vanguard third stage for a combined total thrust of 28,000 pounds. *Vanguard 1* continued to orbit and transmit data for seven years. Its most important discovery was that the Earth was slightly pear-shaped with the narrow end toward the North Pole. The *Vanguard* program continued with other satellites through 1959.

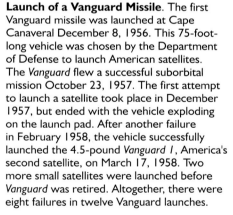

Vanguard 1

The Naval Research Laboratory launched the second U.S. satellite, the *Vanguard 1*, at Cape Canaveral March 17, 1958. This 4.5-pound sphere, six inches in diameter, orbited between 400 and 2,480 miles above Earth. The *Vanguard 1* revealed that our planet is not perfectly spherical, but slightly pear-shaped. Since *Vanguard 1* orbits so high, there is very little air resistance to slow it down; it is expected to remain in space for many years.

Moon — the side that had never been seen by mankind, because the Moon keeps one face turned permanently towards Earth. *Luna 3* was equipped with a very complex onboard imaging system; at the heart were two cameras. The first camera took shots of the entire Moon, and the second took closer shots of selected areas. The satellite reached the far side of the Moon on October 6. The cameras made twenty-nine exposures and transmitted them to Earth, just before making a soft landing on the far side of the Moon. The images were of poor quality by modern standards, but Soviet scientists were able to view the far side of the Moon — a sight never before seen by a human being.

More Soviet Space Probes

From 1963 to 1965, the *Luna 4* through *Luna 8* probes either achieved a fly-by or impacted the Moon.

Luna 9, launched January 31, 1966, carried an egg-shaped container with scientific measuring equipment, radio communications equipment, and a lightweight, power-efficient camera. On February 3, the *Luna 9* space probe landed on the Moon, and within fifteen minutes, began taking pictures of the lunar surface and a panorama of the surrounding landscape. The pictures were of poor quality, but clearly showed rocks and craters.

The Soviet Union continued to send probes to the Moon, *Luna 10* through *Luna 24*. The last of the Luna probes, *Luna 24*, landed on the Moon on August 18, 1976, and collected Moon dust. It returned to Earth on August 23 with samples of small pebbles of iron-rich material about 1/3" in diameter.

Luna Probe Results

The Soviet Luna program produced much valuable information about the Moon's chemical composition, gravity, temperature, and radiation characteristics. Much of this information was shared with U.S. and British scientists.

American Moon Probes

The United States Moon program started by launching nine Pioneer spacecraft missions from 1958 to 1960. None of the Pioneer Moon missions were successful; however, Pioneers 6 through 9, which were successfully launched into solar orbit between 1965 and 1968, are still in solar orbit, and numbers 6 and 8 are still transmitting information. During 1961-1965, nine Ranger spacecraft

Early Pictures from the Moon. The Soviet Union's *Luna 9* was the first space probe to make a soft landing on the Moon and then transmit photographs to Earth. The successful lunar landing on February 3, 1966, ended the speculation that any landing object would sink into the regolith (the layer of fine dust that covers almost the entire surface). This Magyar (Hungary) stamp illustrates the probes landing.

returned over 15,000 photographs of the Moon's surface. During 1966-1968, five Surveyor probes made soft landings on the Moon; *Surveyor 2* and *4* crashed on the Moon's surface.

Surveyor Space Probes

The *Surveyor 1* space probe landed on the Moon June 7, 1966. It carried two television cameras and over one hundred sensors. It landed in the Oceanus Procellarum and succeeded in sending back 11,000 pictures and much data about the surrounding surface.

On April 17, 1967, *Surveyor 3* landed on the Moon in the Mar Cognitum area of Oceanus Procellarum and sent back over 6,300 images.

Surveyor 5, landing on September 8, 1967, in the Mare Tranquilitatis, transmitted nearly 20,000 images and hours of data about the lunar surface. *Surveyor 6*, launched November 7, 1967, landed in the Sinus Medii, and sent a record 30,027 images to Earth.

Surveyor 7, the last probe of the Surveyor program, launched on January 7, 1968, and sent over 21,000 images back to Earth.

All of the Surveyor probes or their remains are still on the Moon.

Other Space Probes

In 1966 and 1967, five Lunar Orbiter probes were sent to take extensive photographic coverage of the Moon. The orbiters covered ninety-nine percent of the Moon's surface.

The Apollo missions were designed to put a man on the Moon and to conduct scientific experiments to investigate the Moon's geology and history. Six Apollo vehicles landed on the Moon and twelve astronauts explored the surface of the Moon using various scientific instruments. Many photographs, test results, and rock samples were returned to Earth.

More Sputniks Launched

In May 1960 the Soviet Union launched *Sputnik 4*, which turned out to be a test flight of the Vostok spacecraft in which Soviet cosmonauts would eventually fly. This flight eventually burned up in the atmosphere.

In 1960, *Sputnik 5* carried two dogs, Belka and Strelka, into orbit for one day, returning them safely to Earth to become canine celebrities as the first living things to return from Earth orbit.

Sputnik 6 carried another two dogs, but they perished when the space capsule entered the atmosphere at the wrong angle and burned up.

SOVIET UNION LAUNCHES FIRST HUMAN TO SPACE

On April 12, 1961, 27-year-old military pilot, Yuri Alekseyevich Gagarin, was rocketed into orbit from the Soviet space base at Tyuratam. After completing one orbit, the automatic controls of his *Vostok 1* spacecraft brought him back to Earth as scheduled. His entire flight, from lift-off to touchdown, lasted 108 minutes. Gagarin became the first man in space, the first to orbit the Earth, and, as a result, was a world hero. In 1968, while training for a *Soyuz* spacecraft mission, he was killed in an aircraft crash.

On August 6, 1961, cosmonaut Gherman S. Titov made a seventeen-orbit flight lasting just over a day. Tivov ate, worked, and slept in orbit. His flight in the *Vostok 2* spacecraft showed that man could survive more than fleeting bouts of weightlessness. Titov's flight stayed aloft for more than twenty-five hours.

First Human Spaceflight. On April 12, 1961, the Soviet Union's cosmonaut Yuri A. Gagarin, aboard a *Vostok 1* spacecraft, became the first Soviet man in space. *Vostok 1* completed one orbit of the Earth and the mission lasted for 108 minutes. Upon reentry Gagarin ejected at an altitude of 4.3 miles. He parachuted separately to the ground. This stamp was issued in Mongolia.

More Soviet Union Records

The Soviet Union launched the world's first "group space flight" in August 1962. Cosmonaut Andrian Nikolayev blasted off in a *Vostok 3* spacecraft. One day later, cosmonaut Papel Popovich blasted into orbit aboard a *Vostok 4* spacecraft. The two spacecraft circled the Earth for several days, not flying together, but within 3.1-miles from each other during one orbit.

A second "group flight" occurred in June 1963, when Valery Bykovsky rode the *Vostak 5* spacecraft into orbit. Two days later, *Vostak 6* was launched into orbit. Again, the two spacecraft did not travel together, but on one orbit, the *Vostak 6* came within three miles of Bykovsky. Cosmonaut Bykovsky in *Vostak 5* posted a record journey of eighty-one orbits.

However, it was the *Vostok 6* that captured all the headlines. Riding in the spacecraft was a woman cosmonaut, 26-year-old Valentina V. Tereshkova. To her went the honor of being the first woman ever to travel in space. Cosmonaut Tereshkova circled Earth forty-eight times.

The First Spacewoman. On June 16, 1963, the *Vostok 6* was launched by the Soviet Union carrying Valentina V. Tereshkova, who became the first woman in space. She made forty-eight orbits of Earth before returning safely on June 19, thus showing that women were as capable of traveling in space as men. Valentina never made another flight. In late 1963, she married fellow cosmonaut Andrian Nikolayev, pilot of the *Vostok 3* satellite, later giving birth to a healthy girl, thus dispelling fears that radiation in space might cause genetic damage. Many other spacemen have subsequently sired perfectly normal children. This Air Mail stamp of *Vostok 6* and Valentina was issued in Mongolia.

NASA IS FORMED

In 1958, Congress established the National Aeronautics and Space Administration (NASA) to conduct the nation's program for the peaceful exploration of outer space. Since then, NASA has become a prime user of the Atlantic Missile Range that stretches from Cape Canaveral through the Bahamas, past the eastern tip of South America, to Ascension Island 5,000 miles away, and the Indian Ocean, 10,000 miles distant.

An announcement by President John F. Kennedy on May 25, 1961, that the United States would send men to the Moon and return them safely to Earth by the close of the 1960s forced NASA to determine a launch method and construct its new launch facilities as soon as possible. It wasn't long before NASA purchased land on Merritt Island, north of Cape Canaveral, selected the *Saturn V* concept as the Moon launch vehicle, selected the first astronauts, and started construction of the Vehicle Assembly Building, Launch Pad 39, crawler/transporter, launch control center, and a sprawling industrial support area.

Meanwhile, the Marshall Space Flight Center in Huntsville, Alabama, was selected to manage construction of the *Saturn V* rocket, the Manned Spacecraft Center in Houston, Texas, was selected to manage construction of the spacecraft, and the Launch Operations Center at Cape Canaveral was selected to manage overall integration, testing, and launch. After President Kennedy's assassination in 1963, the Launch Operations Center was renamed in his honor, becoming the John F. Kennedy Space Center.

The U.S. Air Force continued to manage the Atlantic Missile Range, with NASA designated a user. NASA agreed to expand its land purchases by forty square miles to provide enough land for future Air Force expansion. The Air Force decided to construct a huge launch facility for its *Titan III* rockets on land dredged from the Banana River, north of Cape Canaveral.

In 1963, NASA negotiated land use agreements for submerged lands, and the National Wildlife Service was authorized to administer lands not needed for development. This resulted in the creation of the Merritt Island National Wildlife Refuge. NASA's land acquisition, totaling about 88,000 acres, was completed in early 1964.

Once NASA got down to work, it put an end to the "race" between the three military services. It then began to launch its own unmanned space shuttles and started on a program to send men into space.

NASA's manned space flight program started with Project Mercury, with a goal of placing an astronaut in orbital flight around the Earth, investigating his capabilities while in orbit, and recovering him safely.

Mercury was seen as a necessary prelude to more extensive manned space flights. Project Mercury is discussed further in "Chapter Four."

After completion of Project Mercury, NASA turned to the Gemini program. This program developed the skills and assembled the flight crews, support personnel, and facilities that would be needed to send men to the Moon. A major contribution of the Gemini program was the development and proving of space rendezvous techniques. Rendezvous were achieved under manual control by the pilots, and semi-automatically by the spacecraft's' systems. They were made from above and below the target vehicles. In short, Gemini provided the experience in orbital rendezvous that had been lacking. Gemini also provided the long-duration experience, operational flight control procedures, ground crew skills, precise reentry guidance and landing techniques, extra-vehicular activities and, most importantly, highly trained and experienced flight crews that would bring the Apollo program to its ultimate goal. The Gemini Program is covered more in "Chapter Five."

After completing the Gemini program, NASA started the Apollo program and began to build toward a Moon landing. The first test of the giant *Saturn V* booster rocket, with all its

LEFT:
President John F. Kennedy Visits Cape Canaveral. In early May, Russian Premier Nikita Khrushchev's declaration that Alan Shepard's trip into space was a "flea jump" prompted President John F. Kennedy to plead before a joint session of Congress on May 25, 1961, to step up the effort to put the United States first in the space race. "I believe," he said, "that this nation should commit itself to achieving the goal, before this decade is out, of landing a man on the moon and returning him safely to Earth." The Space Race now became the Moon race. This view shows President Kennedy being briefed by astronaut Walter Schirra at Launch Complex 14 when he visited The Cape on January 1, 1962.

TOP ROW:
Presidential Visit, September 11, 1962. President John F. Kennedy being briefed about the plans of NASA to get a man to the Moon. President Kennedy toured the Cape Canaveral launch facilities.

Presidential Visit, November 16, 1963. Dr. Kurt Debus, director of the Launch Operation Center, greeted President John F. Kennedy and his team on their arrival at Cape Canaveral for a tour of the area.

President Lyndon B. Johnson Visits KSC. Rocco A. Petrone (left), director of KSC Launch Operations, briefs President Johnson (seated at right) and others during the President's September 15, 1964 visit.

stages live, occurred November 9, 1967. An unmanned test of the Apollo spacecraft and lunar lander in Earth orbit followed on April 4, 1968, and the first manned orbital test of the spacecraft, *Apollo 7*, on October 11th of the same year.

Apollo 8 became the first manned spacecraft to reach the Moon in December 1968, when it made ten orbits before returning to Earth. *Apollo 9* astronauts tested the spacecraft and lunar lander in Earth orbit during March 1969, and *Apollo 10* carried out a rehearsal for the docking maneuvers in Moon orbit in May 1969.

On July 20, 1969, astronaut Neil A. Armstrong became the first man to walk on the Moon. Five more landings followed. With *Apollo 17* in December 1972, the lunar exploration program came to an end. The Apollo program is discussed in "Chapters Six and Seven."

AMERICA'S FIRST ASTRONAUTS

Beginning in early 1959, NASA searched for the men who would be its first astronauts. It tested 110 military flyers for the job. Seven men, all jet pilots, were finally chosen. Three came from the Air Force: Captain Leroy Gordon Cooper, Captain Virgil "Gus" Grissom, and Captain Donald "Deke" Slayton; three were Navy pilots: Lt. Malcolm Scott Carpenter, Lt. Commander Walter "Wally" Schirra, and Lt. Commander Alan Shepard; and one was a Marine Corps flier: Lt. Colonel John Glenn.

These seven men immediately became national heroes and their NASA activities made headlines throughout the country. The men went through an intensive physical training program, rode in whirling centri-

The First Seven. The original seven astronauts selected by NASA are shown standing beside a Convair 106-B aircraft. They are, left to right, M. Scott Carpenter, L. Gordon Cooper Jr., John H. Glenn Jr., Virgil I. Grissom, Walter M. Schirra Jr., Alan B. Shepard Jr., and Donald K. Slayton.

The Original Astronauts. This photograph was taken at the Manned Spacecraft Center (MSC) in Houston, Texas. Astronaut Gus Grissom died in the *Apollo 1 — Apollo/Saturn (AS-204)* — fire at The Cape on January 27, 1967. Astronaut Deke Slayton died from complications of a brain tumor in League City, Texas, on June 13, 1993. Astronaut Alan Shepard died after a lengthy illness in Monterey, California, on July 21, 1998. As of January 1, 1977, none of the original seven astronauts remained with the NASA Space Program. However, in October 1998, John Glenn, now a United States Senator (Democrat-Ohio), flew as payload specialist on the Space Shuttle STS-95 mission.

Signature Photograph. The seven original astronauts are pictured around a table admiring an Atlas missile model. Standing, left to right, are Alan B. Shepard Jr., Walter M. Shirra Jr., and John H. Glenn Jr.; sitting, left to right, are Virgil I. Grissom, M. Scott Carpenter, Donald Slayton, and L. Gordon Cooper Jr.

fuges to accustom them to the heavy pressures of gravity at lift-off, floated about in special chambers to prepare them for work when weightless, and were made to endure great heat and high noise levels.

Kennedy Space Center, Florida. The Kennedy Space Center (KSC) was the launch site for the *Apollo* and *Skylab* missions. It is currently responsible for processing, launching, and landing space shuttles and its payloads, including *International Space Station (ISS)* components. This view of Cape Canaveral and the KSC illustrates how The Cape juts out into the Atlantic Ocean. The launch sites for *Mercury* and *Gemini* were on Cape Canaveral.

KENNEDY SPACE CENTER

The Kennedy Space Center is where the dream of human space flight is turned into reality for the people of the United States and many other nations of planet Earth. The center covers about 88,000 acres of land, swamp, and waterways. It is home to two launch pads (Launch Pads 39A and 39B), one of the world's largest runways, the nation's third biggest building (Vehicle Assembly Building), the Visitor Complex, and an administration/industrial section.

KENNEDY SPACE CENTER, FLORIDA

Vehicle Assembly Building (VAB)

Inside the giant caverns of the Vehicle Assembly Building *Saturn V* and Space Shuttle vehicles were assembled on a Mobile Launcher Platform and connections were tested before they were moved out to either Launch Pads 39A or 39B. The process of getting a space vehicle ready for launch begins when an empty launch platform is moved into place in one of the high bays where the work is done. Then, segment-by-segment, the vehicle is assembled on the platform.

Standing 525-feet-tall, covering eight acres, and enclosing a volume of 129,428,000 cubic feet, the VAB is the third largest building in the world and can be seen on the horizon for miles around.

Originally designed to support the stacking of four *Saturn V* moon rockets at the same time, the VAB remains the most visible icon of the Kennedy Space Center and of Florida's Space Coast. Since the 1980s the VAB has been used to assemble the Space Shuttle.

Vehicle Assembly Building (VAB). The *Apollo/Saturn V* facilities vehicle moves out of the Vehicle Assembly Building (VAB) at the Kennedy Space Center, on its way to Launch Complex 39.

Symbols of Supremacy. The magnificent American Bald Eagle shares residence at the Kennedy Space Center with mammoth launch facilities, including the Vehicle Assembly Building (VAB) on the left.

Crawler Transporter. At Launch Complex 39, service engineers check one of the double-tracked crawlers of the six million-pound transporter for the *Saturn V* launch vehicle. After the *Apollo-Saturn V* missions ended in the 1970s, the Crawler Transporter was modified to transport the Space Shuttle from the Vehicle Assembly Building to Launch Pads 39A and 39B.

Launch Complex 39. Launch Pad 39 was built during the 1960s to launch the *Saturn V* rocket. The *Saturn V* was assembled in the Vehicle Assembly Building, shown in the top left corner, and moved to the launch area by a Crawler Transporter. During the 1980s and beyond, Pad 39 was used to launch the Space Shuttle.

Launch Complex 39

Launch Complex 39 was constructed during the mid-1960s as the launching pad for sending men to the Moon atop the *Saturn V* rocket. The Apollo launch facility was later modified so that it could be used to launch the Space Shuttle.

The Visitor Complex

The idea for a Visitor Complex began in the early 1960s, when space program employees and their families were allowed on Sundays to drive through the restricted U.S. Air Force grounds at Cape Canaveral.

This drive-through tour expanded to include the Kennedy Space Center in the mid-1960s. In 1967 the Visitor Complex opened and visitors could sit through a NASA movie, see space hardware, take a tour of launching pads, or purchase space memorabilia from the gift shop.

By 1990, the annual attendance grew to over three million. The KSC Visitor Complex quickly became one of the top tourist attractions in Florida, which resulted in an expansion of facilities.

KSC Visitor Complex. This aerial view of the Visitor Complex shows the huge Vehicle Assembly Building in the background. In the 1990s the Visitor complex was greatly expanded to accommodate the millions of people who wanted to see where NASA sent humans and space vehicles into space.

Space Hardware on Display. Space travel hardware on display at the KSC Visitor Complex: Apollo Command Module, Explorer I Satellite, and outside the building, the Moon landing vehicle.

Gemini IX Spacecraft. On display at the Visitor Complex is the actual *Gemini IX* spacecraft flown by astronauts Thomas Stafford and Gene Cernan on June 3-6, 1966. *Gemini IX* spent more than seventy-two hours in Earth orbit, including a 125-minute spacewalk by Cernan.

⊕ The Space Age Begins

Today, the KSC Visitor Complex includes many separate exhibition, administration, and concession buildings, two IMAX theaters with five-story screens, a Rocket Garden, the Apollo/Saturn V Center, International Space Station Center, Shuttle Launch Simulator, Astronaut Memorial, Full Scale Shuttle replica, space robot exhibit, U.S. Astronaut Hall of Fame, huge gift shop, and a fleet of forty-four air-conditioned tour busses.

Welcome to the Visitor Complex. This 1970s view shows the front entrance of the KSC Visitor Complex, where several missiles were on display.

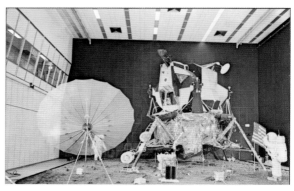

Lunar Surface Diorama. A 1970s Moon landing exhibit at the KSC Visitor Complex.

Space Shuttle Exhibit at KSC. This Space Shuttle may look and even feel real, but it is actually a full-size replica of an actual shuttle. At the KSC Visitor Complex, guests can experience what it feels like to walk and work aboard a real Space Shuttle within this *Explorer* shuttle replica.

Impressive Spacecraft Display at KSC. Shown is a *Skylab* docking adapter, Apollo spacecraft, and hanging from above, a Russian *Soyuz* spacecraft.

Space Shuttle Fuel Tanks. The huge liquid hydrogen/oxygen main Space Shuttle tank and solid rocket fuel boosters are on display at the Visitor Complex.

Rocket Garden. In the Rocket Garden at the KSC Visitor Complex, visitors can see the rockets that launched astronauts and satellites into space including a Mercury Redstone, similar to the one that carried astronaut Alan Shepard, the first American to venture into space. Another special vehicle in the collection is an awe-inspiring Mercury Atlas identical to the rocket that carried astronaut John Glenn into space for America's first orbit of Earth.

Apollo/Saturn Visitor Center

The centerpiece of the Apollo/Saturn V Center is a fully-restored *Saturn V* Moon rocket. The *Saturn V* rocket, which for twenty years rested near the Vehicle Assembly Building exposed to the corrosive sea air, is made up of stages destined for Apollo missions that were cancelled. It is one of three remaining in the world. Visitors and tourists can walk around and under this enormous Moon rocket.

Also exhibited are an original Lunar Excursion Module and Command Service Module. The Command Module was used for the Apollo/Soyuz Test Project in 1975. A variety of other related items, including a "moon rock" that you can touch, are on exhibit.

Apollo/Saturn V Center. This aerial view shows the Apollo/Saturn V Center, which is part of the KSC Visitor Complex. The long building in the center contains a full size *Saturn V* rocket — the vehicle that took astronauts to the Moon.

Saturn V Moon Rocket. Shown is an actual 363-foot-long, 6.2-million-pound *Saturn V* Moon rocket, with a dramatic recreation of the first manned Moon landing, and hands-on exhibits. Visitors can actually touch a "Moon rock." The *Apollo/Saturn V* Center brings to life the U.S. space program's mission to the Moon.

The International Space Station Center

The three-story Space Station Center is a 457,000-square-foot building that is used to prepare parts for the *International Space Station*. When ready-to-go these parts or modules are carried into space in the cargo bay of the Space Shuttle. This Space Station processing facility includes two bays: one where the hardware can be worked on in a "clean" environment and one where visitors can explore actual Space Station hardware. Visitors can view modules as they are prepared for launch.

Launch Control Center

At the Launch Control Center, the KSC launch team gathers to monitor and direct all efforts to prepare a space vehicle for launch. It is home to the computers that control the last nine minutes of the count down. After a successful launch, vehicle control is transferred to the Johnson Space Center in Houston, Texas.

Crawler Transporters

The Crawler Transporter moves the Mobile Launcher Platform, containing an assembled space vehicle, from the Vehicle Assembly Building (VAB) to one of the launch pads, 39A or 39B. The Crawlers maximum speed is two miles per hour. The normal speed of a Crawler Transporter is about one-mile per hour while transporting a space vehicle. The Crawler travels on a special designed highway between the two launch pads and the Vehicle Assembly Building.

The Crawler Transporter is about half the size of a soccer field and weighs more than six million pounds.

Shuttle Landing Facility

The Shuttle Landing Facility is one of the longest and widest runways in the world. The runway, which can be seen from space, is 15,000 feet long, three hundred feet wide, 1,000 feet of paved overruns at each end, and an average of sixteen inches thick. At night, airplane pilots say they can see its lights from as far away as Jacksonville, 160 miles to the north.

U.S. ASTRONAUT HALL OF FAME

The United States Astronaut Hall of Fame, located on State Road 405, six miles west of the KSC Visitor Complex, exhibits the largest collection of personal astronaut memorabilia ever assembled. Visitors can explore a rare collection of astronaut artifacts, displays, and interactive exhibits dedicated to honoring astronauts. Visitors can also ride a Rover across the rocky Mars terrain and blast off into an interactive experience that gives a true taste of space. ATX (Astronaut Training Experience) is an intense half-day of hands-on training and preparation for the rigors of spaceflight. ATX train-

International Space Station Center. At this facility, visitors can see *International Space Station* modules as they are prepared for launch.

U.S. Astronaut Hall of Fame. Located directly across the Indian River from the Kennedy Space Center, the U.S. Astronaut Hall of Fame has the largest collection of personal astronaut memorabilia ever assembled, as well as many interactive exhibits.

ing includes operating a full-scale Space Shuttle mock-up and taking the helm in mission control. A life-size statue of astronaut Alan Shepard, the first American that was hurled into space, is located in the entrance lobby.

The KSC Buildings

Located in the KSC Industrial Area, the four-story Headquarters Building is where KSC's top managers and support staff work. This building also includes a library, cafeteria, post office, and an employee gift shop.

The four-story Operations & Checkout Building, built during the 1960s for the Apollo program, is today used to support the Space Shuttle program. During the Apollo program, the four-story-tall high bay was used to check out the astronaut's spacecraft. The O & C Building is also famous because astronauts who are about to depart on a Space Shuttle mission stay there during the days just prior to launch. Many historic photographs have been taken capturing the moment when a mission crew departs the O & C Building for the launch pad.

Several other buildings in the Industrial Area are used for various purposes: hazardous work, repair work, satellite processing, solid rocket booster parachute checkout, and fueling probe rockets.

MERRITT ISLAND NATIONAL WILDLIFE REFUGE

Mention Canaveral and Merritt Island and many people imagine space exploration — the excitement of the countdown and the thrill of liftoff. However, there is another kind of space here, one that is almost a secret from outsiders. It is called the Merritt Island National Wildlife Refuge, which NASA gave to the U.S. Fish and Wildlife Service in 1963. The area was originally set aside in the 1950s as a buffer zone for nearby NASA activities. The 140,000-acre refuge shares a common boundary with the Kennedy Space Center and is the habitat for wildlife and plants, where the warm waters and temperate climate nourish species as diverse as bald eagles, alligators, bromeliads and mangroves. This diverse landscape is home for over 350 species of birds, thirty-two species of mammals, 117 species of fish, sixty-eight species of amphibians and reptiles, and over 1,000 species of plants. Manatees, dolphins, sea turtles, bald eagles, otters, alligators, bobcat, and wading birds live in the shadows of the world famous launch pads where space exploration was born.

CAPE CANAVERAL OR CAPE KENNEDY?

A week after President John F. Kennedy's assassination in 1963, the names of NASA and Air Force facilities in Central Florida, as well as certain geographic features there, were changed to honor the slain President.

As so many historic space missions were launched, one after another, people around the world became familiar with the name Cape Kennedy, but by 1973 public sentiment in Central Florida prompted congress to change some of the names back.

So today NASA launches Space Shuttles from the Kennedy Space Center on Merritt Island while the U.S. Air Force operates the Cape Canaveral Air Force Station, which is located on the geographic feature known as Cape Canaveral.

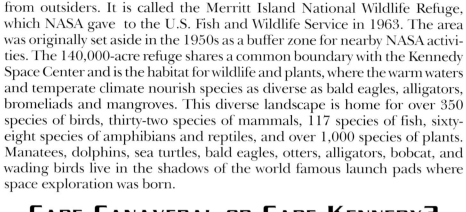

Birds and Rockets. This mother eagle looks at one of her offspring, at left, in their nest at the north end of State Road 3 near Kennedy Space Center. The resident eagle parents are raising two offspring. This year-old nest is one of a dozen eagle nests both in KSC and in the Merritt Island National Wildlife Refuge, which surrounds KSC. The refuge includes several wading bird rookeries, many osprey nests, up to four hundred manatees during the spring, and approximately 2,500 Florida scrub jays. It also is a major wintering area for migratory birds. More than five hundred species of wildlife inhabit the refuge, with fifteen considered federally threatened or endangered.

Merritt Island Wildlife Nature Refuge. On NASA's Kennedy Space Center in Florida, an alligator rests on the bank of a canal. Alligators can be spotted in the drainage canals and other waters surrounding Kennedy. The center shares a boundary with the Merritt Island Wildlife Nature Refuge, which is a habitat for more than 310 species of birds, twenty-five mammals, 117 fishes, and sixty-five amphibians and reptiles.

PROJECT MERCURY

PROJECT MERCURY WAS FORMALLY ANNOUNCED AS THE NAME OF THE FIRST NASA manned space program on December 17, 1958, the fifty-fifth anniversary of the first flights of the Wright brothers. The name of the project was based on a popular name, Mercury, the Roman messenger of the gods.

MERCURY LAUNCH VEHICLES

The Mercury program, planned to get man into space in a minimum amount of time, made use of available systems and components wherever possible. The crew capsule had been designed to weigh about the same amount as a standard warhead for the ballistic missiles then under development, following the reasoning that it was inevitable that those missiles would be used, in slightly modified forms, as the launchers in Project Mercury.

Suborbital Flights

For the suborbital Mercury flights, the Redstone missile was chosen as the launch vehicle. The basic missile was modified by extending its cylindrical tank section to include more fuel for increased cutoff velocity. Simplifications were introduced to increase the potential reliability of the launcher, and an adapter was designed and built to match the Mercury spacecraft. In its final form, the Redstone launch vehicle stood fifty-nine feet high from the base of its fins to the frame that held the spacecraft. The Mercury capsule and escape tower added another twenty-four feet to the height of the assembly.

Orbital Flights

For the orbital flights of Project Mercury, more power was necessary. The U.S. Air Force Atlas ICBM had been tapped for the second phase of the Mercury launches. It had the potential to do the job, and it had Air Force support behind it, determined to make it a reliable operational weapon. The Atlas-D model was chosen and modified. An abort sens-

Mercury-Atlas Missile. For the Earth orbital flights of Project Mercury, NASA selected the General Dynamics' Atlas ICBM. Modified as the Mercury-Atlas launch vehicle, it underwent test launches to qualify its systems for spaceflight. In 1962, it stood on Launch Pad 14 at Cape Canaveral, prepared for an orbital flight.

ing system was developed and installed as additional protection for the astronaut. The Posigrade rockets, used to guarantee separation of the warhead, were removed because Mercury carried its own rockets to aid separation.

The final Atlas launcher configuration resembled the standard missile right up to the part where the warhead would have been mounted. At that point there was an adapter that held the Mercury spacecraft and escape tower. The assembly stood over ninety-five feet tall ready to launch.

THE MERCURY CAPSULE

The one-man Mercury spacecraft was conical in shape, 6.2 feet across at its widest point, 9.5 feet long, and weighed 1.3 tons. Inside the spacecraft, the astronaut sat clad in his spacesuit on a couch contoured to his body shape, his head touching one wall and his feet braced against the other. It was a tight squeeze. In front of him was an instrumental panel, and in his hand he held a joystick with which he could control the orientation, or altitude, of the spacecraft while in orbit. At the spacecraft's blunt end was the heat shield, designed to resist fierce temperatures as the capsule blazed into the atmosphere during reentry. The manned spacecraft reentered blunt end first to slow down before its parachutes opened. Strapped to the back of the Mercury heat shield were the retro-rockets, which fired to put the spacecraft on course back to Earth. The Mercury capsule splashed down in the ocean with the astronaut safely aboard. Parachutes would help break the capsule's fall. The capsule and astronaut would be picked up by helicopters and taken to a nearby Navy ship.

The Mercury capsule was barely big enough for one astronaut. Among the astronaut's jet pilot friends, the capsule earned the nickname "man in a can." A porthole on one side of the capsule and a periscope overhead provided a very limited view outside the spacecraft.

The outside walls of the Mercury capsule were constructed of nickel alloy with an outer skin of heat-resident titanium.

MERCURY ASTRONAUTS

Seven astronauts were selected to fly the Project Mercury missions. Senior in age and date of rank was Marine Lieutenant Colonel John H. Glenn, Jr.; Lieutenant Commanders Walter M. Schirra, Jr. and Alan B. Shepard, Jr., and Lieutenant Malcolm Scott Carpenter came from the U.S. Navy; Captains Donald K. Slayton, Leroy Gordon Cooper, Jr., and Virgil I. Grissom were from the Air Force. They were the first of America's astronauts.

The Mercury Astronauts. These seven men, wearing space suits in this portrait, composed the first group of astronauts announced by the National Aeronautics and Space Administration (NASA). They were selected in April 1959 for the Project Mercury program. Front row, left to right, are Walter M. Schirra Jr., Donald K. Slayton, John H. Glenn, Jr., and M. Scott Carpenter. Back row, left to right, are Alan B. Shepard Jr., Virgil I. Grissom and L. Gordon Cooper Jr.

Astronaut Exhibit. This exhibit of the original seven American astronauts appears in the Wax Museum in Denver, Colorado.

One of the objectives of Project Mercury was to find out what man's capabilities were in space flight. There was plenty for him to do aboard the spacecraft. He could be a useful communications link; he could change the altitude of the spacecraft; in the event of failure of any automatic system, he would be able to take control and to complete the mission.

The importance of keeping man in the loop was followed through in the engineering design of the Mercury spacecraft. Manual control capability was built into the spacecraft, giving the astronaut pilot an override of many automatic features. This paid off later in the Mercury program. In each one of the orbital flights, there was some kind of system malfunction that could have compromised the mission.

Astronaut John H. Glenn, in the MA-6 mission, was plagued with three malfunctions, one of them being the failure of the control system thrusters. He was able to fly manually to complete the planned three orbits. If the astronaut had not been in the control loop, the mission would have been aborted after one orbit, but potentially the most dangerous was a heat shield problem that resulted in Glenn overriding the automatic features and the mission was a success.

Astronauts Walter Schirra in MA-8 and Leroy Gordon Cooper in MA-9 also were able to override an automatic system that had malfunctioned. Schirra's problem was in his suit, which started to show an increase in temperature from the automatically reduced flow of coolant. Manual control of the coolant valve corrected the problem. Cooper was able to fly his spacecraft manually during retrofire and reentry after the automatic control system developed short circuits and malfunctioned.

The Mercury astronauts wore a pressure suit that would inflate and sustain them in an emergency situation, such as a loss of pressure in the spacecraft cabin.

Space Medicine

With men on board the spacecraft, NASA turned their attention towards the disciplines of space medicine. Space flights were known to present new hazards to human life, but the degree of danger was not well defined. Weightlessness was a big unknown. Radiation was thought to be a possible problem. The environmental stresses of vibration, g-loads, noise, temperature, and other characteristics of launch and space flight were also thought to be a problem. The astronauts would be under mental stresses that were new as a result of their environment. Loneliness might be a major complaint while noxious or toxic fumes at best would be annoying and at worst might imperil the astronauts.

NASA felt the answers to these questions would hold the key to all future space exploration. In 1960, it established its first Office of Life Sciences, with the long-term goals related to the exploration of space and the possible effects of the space environment on biology.

NASA doctors developed monitoring sensors that were attached to the astronauts. Breathing rate, heart rate and rhythm, blood pressure, and body temperature were the basic parameters monitored during the Mercury flights. The experience gained in medical planning and monitoring during all phases of Project Mercury proved invaluable and resulted in modified procedures being adapted for future space programs.

Early Spacecraft Tests

On December 19, 1960 an un-piloted Mercury capsule was launched on top of a Redstone missile, which was successfully recovered from a suborbital flight — a flight into space that does not achieve sufficient altitude to orbit Earth.

Another test followed in January 1961. The flight had a passenger: Ham, a three-year-old chimpanzee. Ham's flight lasted only eighteen minutes and he returned to Earth in excellent health. During his short flight, Ham performed all his tasks well and his flight helped NASA know how a person might function in space and proved that space travel was safe.

In May 1961, an unmanned Mercury capsule went up on an Atlas rocket test launch. A critical failure occurred and, hidden inside a cloud, the Atlas exploded. The capsule soon appeared, drifting back safely, indicating that the launch escape tower pulled it free.

In November 1961, a Mercury-Atlas vehicle was launched at Cape Canaveral carrying chimpanzee Enos into space on a successful two-orbit flight.

Chimpanzees Make Space Flights. Before human astronauts were sent into space, two chimpanzees, named Ham and Enos, made space flights in Mercury spacecraft tests.

FIRST AMERICAN IN SPACE

The nation's eyes were on NASA's next launch. Some 350 media correspondents busily streamed out their reports from The Cape. Alan Shepard's astronaut comrades were all around him, involved in every part of the operation. John Glenn had breakfast with him and stood in for him during pre-launch preparations. Deke Slayton was on the radio link in Mercury Control nearby. Gordon Cooper was in the blockhouse with the launch crew. Wally Schirra was circling in his F-106 chase plane, scouting the winds and ready to follow Shepard as high as he could.

Shepard, settling inside his spacecraft atop a Redstone launch vehicle, told the launch crew, over the communications loop, "Let's light this candle." The firing button was pushed and seconds later Shepard was streaking through the sky at 5,100 miles per hour.

America's first manned space flight was a suborbital journey made by astronaut Alan B. Shepard, Jr., in *Freedom 7*. This Project Mercury flight, launched at 9:43 a.m., on May 5, 1961, on Launch Pad 5 at Cape Canaveral, climaxed the years of work that had preceded the successful flight and marked the beginning of America's manned ventures into the space environment.

Shepard was lobbed 302 miles into the Atlantic Ocean from Cape Canaveral, reaching a maximum altitude of 116.5 miles. He was able to make observations of Earth and the sky and control the altitude of his spacecraft with tiny rocket jets.

At 21,000 feet above Earth, a small parachute opened to steady the fall, followed by the main parachute. The spacecraft, with Shepard inside, plunged into the Atlantic Ocean. It landed just forty miles from the planned target. Within minutes, Shepard was safely aboard the recovery ship, the U.S. Navy aircraft carrier *U.S.S. Champlain*.

Returning the spacecraft to Earth required the use of retro-rockets, a heat shield, and parachutes. Fired in the direction of the moving craft, retro-rockets started the slowing process. As the craft lunged through the atmosphere, the heat shield protected the craft and astronaut from the intense heat caused by friction. A series of deployed parachutes allowed the craft to "splashdown" into the ocean and be picked up and transported to the recovery ship.

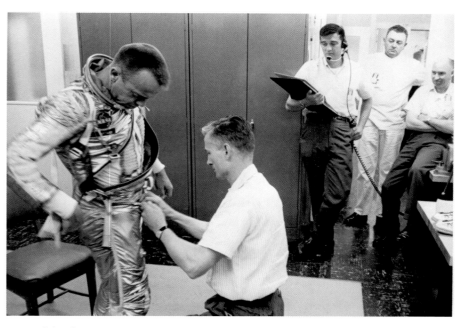

TOP, RIGHT: Getting Ready to Go. Astronaut Alan B. Shepard is being assisted getting into his space suit by a technician at 2:45 a.m. in the morning, beginning the long countdown of the scheduled launch of the Mercury Redstone. After he suited up, Shepard traveled by van to the Redstone Gantry on Pad 5 and was placed into the Mercury Capsule, called *Freedom 7*, on top of the Redstone booster rocket. Shepard piloted the first U.S. manned suborbital space flight of the Manned Project Mercury program.

With the exception of a few minor problems, the fifteen-minute, 22-second flight proved to be a success. Shepard confirmed what Soviet cosmonaut Yuri Gagarin had shown: that there was no obstacle to man functioning successfully in space.

Alan Shepard's pioneering effort did more than put an American in space; it made him a national hero. Shepard's success prompted President John F. Kennedy to go before a joint session of U.S. Congress to pledge his commitment to carry out one of mankind's oldest dreams. On May 25, 1961, less than three weeks after Shepard's flight, the President delivered these immortal words:

I believe that this nation should commit itself to achieving the goal before this decade is out, of landing a man on the Moon and returning him safely to Earth. No single space project will be more exciting or more impressive to mankind, or more important for long range exploration of space, and none will be so difficult or expensive to accomplish.

It was a challenging task to ask of the nation, when U.S. morale was low, despite Alan Shepard's suborbital flight. The fact remained that the Soviet Union had orbited a cosmonaut around the world while America had gone only a small part of the way. However, many NASA officials were convinced that they could meet President Kennedy's goal by the end of the 1960s. They immediately set to work and started building a team to perform the task.

First American in Space. In 1961, the U.S. space
program was lagging behind that of the Soviet Union.
On April 12, Soviet cosmonaut Yuri Gagarin had not
only already penetrated the barrier of space but
had actually orbited the Earth. Manned flight did not
become a reality in America until May 5, 1961, when
Alan B. Shepard Jr. blasted off in a Mercury-Redstone
rocket, *Freedom 7* spacecraft, to become the first
American in space. In a sixteen-minute ride, he
ascended 115 nautical miles, covering a 302-mile range.

Recovering Shepard from His Spacecraft.
A Marine helicopter recovery team hoists astronaut
Alan Shepard from his *Freedom 7* spacecraft after
a successful suborbital flight and splashdown. His
spacecraft plunged into the Atlantic Ocean, forty
miles from the planned landing site, where he and
his spacecraft were recovered by helicopter and
transported to the awaiting U.S. Navy aircraft
carrier *U.S.S. Champlain*. As he stepped on the
carrier deck, he stated, "What a ride!"

End of a Successful Flight. The crew of the U.S. Navy
carrier *U.S.S. Champlain* cheer and take pictures of the arrival
of the first Project Mercury pilot to fly a suborbital flight,
Astronaut Alan B. Shepard Jr. Two marine helicopters are
approaching the aircraft carrier, one carrying the astronaut;
the other, the *Freedom 7* capsule.

SECOND SUBORBITAL FLIGHT

In July 1961, astronaut Virgil I. Grissom repeated the general pattern of Alan Shepard's flight on the second suborbital launch, planned to check the functions of the spacecraft and to evaluate its operation. Grissom's flight nearly ended in disaster when the hatch of his spacecraft accidentally blew off after splashdown. The spacecraft sank, but fortunately the Marine Corps recovery helicopter rescued Grissom. *Liberty Bell 7* sank to the ocean floor. The water it had taken on made it too heavy for the helicopter to pull it out.

In 1999, Grissom's Mercury capsule was recovered from a depth of three miles on the ocean floor in remarkably good condition.

Grissom's flight was the last Mercury mission that used the Redstone launch vehicle. It was not powerful enough to carry the Mercury space-craft into orbit. For the next stage of Project Mercury, NASA prepared to launch men atop a more powerful Atlas booster. The Atlas, America's first intercontinental ballistic missile (ICBM), was built by General Dynamics' Convair Division for the U.S. Air Force. This was a long-range missile with the power to carry a lightweight atomic bomb all the way to the Soviet Union. In 1962, the Atlas was the only U.S. rocket available that had a chance of putting a man in orbit.

Attempt to Recover Liberty Bell 7. Attempted recovery of Mercury spacecraft *Liberty Bell 7* at the end of the Mercury-Redstone 4 (MR-4) mission. Marine helicopter appears to have *Liberty Bell 7* in tow after Virgil I. Grissom's successful flight of 305 miles down the Atlantic Missile Range. Minutes after astronaut Grissom got out of the spacecraft, it sank. The capsule, filled with water, almost pulled the helicopter to the water's surface, and the rescue mission had to be abandoned. In the upper left corner of the view, the recovery ship and another helicopter can be seen.

Convair engineers had developed a lightweight Atlas missile at the expense of reducing structural support. It was a new approach to rocket design; however, the missile had several problems in its early life. The Atlas exhibited great power in getting off the launch pad, but often burst into a spectacular fireball. Four out of ten Atlases exploded after launch.

FIRST AMERICAN TO ORBIT THE EARTH

NASA selected John Glenn to pilot the first manned Atlas flight. His task would take him on a journey that circled the Earth. With this feat, America would catch up to the Soviet Union, at least for a short while.

Astronaut John H. Glenn, Jr. was loaded into his *Friendship 7* spacecraft atop the Atlas booster missile on Launch Pad 14 on February 20, 1962. The countdown hit zero and the Atlas engines ignited. Flame shot out the base of the silver Atlas missile and thunder rocked across The Cape.

Over 50,000 workers gathered at The Cape to watch Glenn take this historic journey into space. Astronaut Glenn reached orbit safely and shot around the Earth at 17,500 miles per hour. His *Friendship 7* spacecraft circled the Earth three times.

Glenn became the first American to see a sunrise and a sunset from space. In fact, he witnessed three sunrises and three sunsets. He found he could distinguish details on Earth with great ease, such as ocean currents, cities, and land formations. He used his hand controls to guide the Mercury capsule during flight, showing a man could safely pilot a spacecraft in orbit. He was also the first Mercury astronaut to eat in space.

Glenn's *Friendship 7* spacecraft landed in an area in the Atlantic Ocean approximately eight hundred miles southeast of Cape Canaveral, in the vicinity of Grand Turk Island. He landed forty-one miles west and nineteen miles north of the planned impact point. The time of the flight — from launch to impact — was four hours, fifty-five minutes, and twenty-three seconds.

America erupted in pandemonium to cheer the country's newest spaceman. John Glenn was a national hero after his historic flight. His flight had restored America's confidence in its space program. It made people believe the United States could indeed compete in technical areas with the Soviet Union. Glenn later met President John F. Kennedy at Cape Canaveral. The wave of patriotism continued on from there. Glenn and his wife Annie were honored by a parade in Washington, D.C., a ticker-tape parade in New York City, and a hometown parade in New Concord, Ohio. Glenn was given the honor of addressing a joint session of Congress a few days after his historic flight. His successful flight made newspaper headlines all over the world.

9-8-7-6-5-4-3-2-1-Fire!
The blast-off of the Mercury *Friendship* 7 spacecraft atop an Atlas booster; the space vehicle is carrying astronaut John H. Glenn, Jr.

Astronaut Glenn Climbing Into His Spacecraft. As the end of the countdown drew near, astronaut John Glenn climbed into the *Friendship* 7 spaceship. He was about to become the first American to orbit Earth.

First American to Orbit the Earth. On February 20, 1962, astronaut John Herschel Glenn Jr., wearing a pressurized suit, almost filled the tiny capsule of Mercury *Friendship* 7 as he prepared for the first American manned flight to orbit the Earth. Traveling 17,500 miles per hour, 160 miles above the Earth, Glenn splashed down safely after orbiting the Earth three times during the five-hour trip.

U. S. Astronaut John H. Glenn, Jr.

Retrieving Glenn's *Friendship 7* Capsule. The Mercury-Atlas 6 *Friendship 7* spacecraft is retrieved from the Atlantic Ocean following astronaut John H. Glenn Jr.'s three-orbit space mission. In this view, the capsule is still in the water, with retrieval cable connected to it.

CARPENTER ORBITS THE EARTH

On May 24, 1962, astronaut Malcom Scott Carpenter flew a flight similar to the John Glenn flight, with a triple orbit of Earth in his *Aurora 7* spacecraft. Technical troubles led to Carpenter overshooting the landing site by 248 miles. Recovery ships had to scramble to reach him; however, he was safely rescued.

Second American to Orbit the Earth. Astronaut Scott Carpenter, dressed in space suit, preparatory to America's second manned orbital flight, May 24, 1962.

America Honors Astronaut Glenn. John and Annie Glenn, with President Lyndon Johnson looking on, in a Washington, D.C. parade.

Retrieving Carpenter's *Aurora 7* Spacecraft. Astronaut Scott Carpenter's *Aurora 7* capsule being lifted out of the ocean by the Navy destroyer *U.S.S. Pierce* after splashdown.

A Six-Orbit Flight

In October 1962, astronaut Walter M. Schirra, Jr., doubled the orbital flight time of his two predecessors in a six-orbit trip around Earth and a duration of more than nine hours in space. Slowly, the experiences were building, with the data being obtained for evaluating the capabilities of man in space.

Frogmen Helping to Retrieve Schirra's *Sigma 7* Spacecraft. Navy frogmen astride Mercury-Atlas 8 capsule, the Sigma 7, deploy a flotation collar to secure towline for recovery by *U.S.S. Kearsarge*.

U. S. Astronaut Walter M. Schirra, Jr.

Astronaut Walter M. Schirra, Jr. U.S. Navy Commander Walter M. Schirra, Jr., was born March 12, 1923, in Hackensack, New Jersey. Schirra was the pilot of Mercury 8 (MA-8) flight of *Sigma 7* spacecraft on October 3, 1962. Launched from Pad 14 at Cape Canaveral, he completed a successful six-orbit mission around the Earth at an orbital velocity of approximately 17,500 miles per hour. Schirra's *Sigma 7* spacecraft made nearly a pinpoint landing as planned, in the vicinity of Midway Island. Time of the flight was nine hours and thirteen minutes.

Houston Welcomes Astronaut Schirra. Astronaut Walter Schirra during a welcome parade in Houston, Texas. He later flew in Gemini and Apollo space missions, becoming the only Mercury astronaut to fly in all three of America's first space programs.

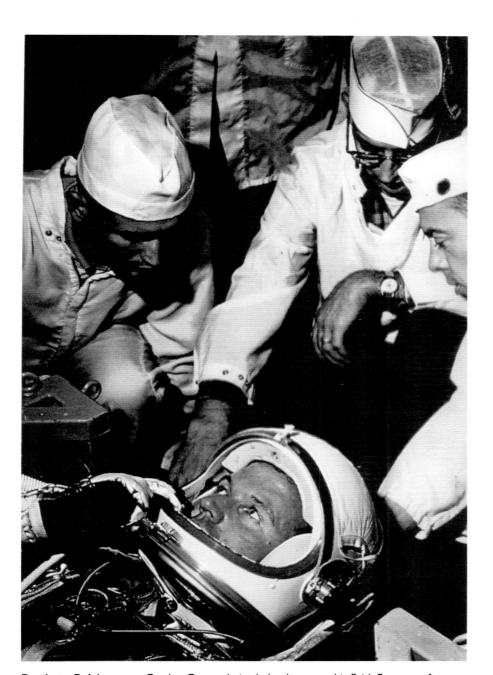

Ready to Go! Astronaut Gordon Cooper being helped to enter his *Faith 7* spacecraft. Astronaut Cooper made the final flight of the Mercury program. He stayed in space more than thirty-four hours and orbited the Earth twenty-two times.

FINAL FLIGHT OF PROJECT MERCURY

It was left to astronaut Leroy Gordon Cooper, Jr., to make the final flight and meet the final objective of the Mercury program. Cooper spent more than a day in space, completing twenty-two orbits of Earth in more than thirty-four hours. Cooper's flight was the most ambitious of all, pushing the capabilities of Mercury to the limit.

During his flight Cooper tried out various sorts of space food. Some food came as bite-sized chunks while other food was a paste in a plastic tube, which could be squeezed out into his mouth. Cooper's flight lasted long enough for him to take a seven-hour nap. He was the first astronaut to sleep in space.

One special feature of Cooper's flight was the observations he made of the ground. Photographs taken by previous astronauts had demonstrated the value of observing Earth and the Atmosphere from space. In addition to continuing this photographic program, Cooper reported seeing astoundingly fine details on the ground by eye, such as roads and railway lines. Later missions confirmed these sightings.

President John F. Kennedy described Cooper's flight in words that could well have applied to the entire Mercury program when he called it "one of the victories of the human spirit."

PROJECT MERCURY RESULTS

The purpose of Project Mercury was to accumulate knowledge about man's capabilities in space for use in future space missions. The flights of Shepard through Cooper had been the first steps — a slow and methodical progression that increased the times in orbit and the amounts of data returned from the flights. The duplicating flights were planned and made to assure that more than luck or coincidence was involved.

Project Mercury succeeded in placing six astronauts into space. The six missions totaled two days, five hours, fifty-five minutes, and twenty-seven seconds of manned space flight. Project Mercury proved that human beings could survive in space and prepared NASA for future missions. It was the United States' first step toward the Moon. The Mercury program paved the way for the Gemini program. NASA and America was ready for a new space program — one with a larger spacecraft and higher goals.

The following table presents a brief summary of the Mercury manned missions:

Mercury Missions

Mercury 3
Date: May 5, 1961
Astronauts: Alan Shepard
Launch Vehicle: Redstone
Earth Orbits: Suborbital

Mercury 4
Date: July 21, 1961
Astronauts: Virgil Grissom
Launch Vehicle: Redstone
Earth Orbits: Suborbital

Mercury 6
Date: February 20, 1962
Astronauts: John Glenn
Launch Vehicle: Atlas
Earth Orbits: Three

Mercury 7
Date: May 24, 1962
Astronauts: Scott Carpenter
Launch Vehicle: Atlas
Earth Orbits: Three

Mercury 8
Date: October 3, 1962
Astronauts: Walter Schirra
Launch Vehicle: Atlas
Earth Orbits: Six

Mercury 9
Date: May 15-16, 1963
Astronauts: Gordon Cooper
Launch Vehicle: Atlas
Earth Orbits: Twenty-Two

Project Mercury Monument. The monument, located at Launch Pad 14 at Cape Canaveral, honors Project Mercury and features the number 7, representing the seven original astronauts, inside the astronomical symbol for the planet Mercury. On November 10, 1964, records and mementos of America's first man-in-space program were placed in the concrete base of this monument, to be opened five hundred years later to provide future generations with accurate historical data on Project Mercury.

GEMINI PROGRAM

O N DECEMBER 7, 1961, NASA ANNOUNCED A SECOND MANNED SPACE program, and it was officially designated the Gemini program January 3, 1962. Named for the third constellation of the Zodiac, Gemini would be a two-man crew. The three main objectives of the Gemini program were: 1. Observe how man and equipment endure space flights lasting up to two weeks; 2. Rendezvous and dock with orbiting vehicles; and 3. Improve reentry procedures at a pre-selected site.

GEMINI LAUNCH VEHICLES

The U.S. Air Force *Titan II* Intercontinental Ballistic Missile (ICBM) was chosen to launch the Gemini spacecraft. The Titan II ICBM carried a much larger payload than the Atlas had carried and could handle the larger Gemini spacecraft. Titan used a propellant combination that could be stored, in contrast to the liquid oxygen in the Redstone and Atlas missiles, which boiled off during lengthy delays in firing and had to continually be topped off to maintain fuel capacity.

The changes to Titan II involved mostly modifications and simplifications rather than construction of any major new components. However, redundant or backup systems were added to improve flight safety. The two stages of the Titan II missile towered ninety feet above the launch pad, and the spacecraft added another nineteen feet.

Launch Complex 19 was converted into the pad for the Titan II missile. Specialized equipment was added at many points to support the addition of the intricate space capsule to the top of the Titan II. The blockhouse got new computers and systems monitors. The leaning service tower had to be extended to accommodate the revised missile, which stood several feet taller with the spacecraft setting on top.

Media at the time generally framed Gemini as a NASA operation, but in truth it was a multilateral team effort — a NASA spacecraft and crew launched on a U.S. Air Force rocket from an Air Force launch pad by a contractor (Martin Company) launch team.

GEMINI SPACESUITS

The basic Gemini spacesuit was a four-layer garment with oxygen ducts for breathing and ventilation. It was used successfully on the short duration missions, but for the fourth Gemini flight, when extra-vehicular activity (EVA) was introduced, the suit had to be modified further.

The EVA suit had added layers of material to protect the astronaut from micro-meteoroid hits while he was outside the spacecraft, and to give him more thermal protection. The helmet had a second visor, intended for use only outside the spacecraft, to give extra protection from the sun and to protect the inner visor from any possible impact damage.

A new long-duration suit was designed for the planned fourteen-day *Gemini VII* mission. It was a "soft" suit, with a fabric hood, which could be removed for storage. Being lightweight, the suit gave a maximum of freedom and comfort to the astronaut.

GEMINI SPACECRAFT

Whereas the Mercury spacecraft had carried only one man and had small jets to control the altitude of the spacecraft, the larger Gemini spacecraft would carry two men and have an onboard propulsion system for orbital maneuvers. The Gemini spacecraft were also equipped with guidance and navigation systems and rendezvous radar, thus enabling the astronauts to try out various rendezvous and docking techniques essential for lunar missions.

The second-generation Gemini spacecraft was modular design. The astronauts occupied the crew module, or reentry capsule. Behind the crew capsule was an equipment module, and behind that a retro-module that housed the sixteen retro-rockets. Before the Gemini capsule returned to Earth, the equipment and retro-modules would be jettisoned. The retro-rockets, or small rocket thrusters, enabled the astronauts to control the attitude of the spacecraft and to change orbits for rendezvous and docking with a target vehicle.

The Gemini capsule had opening hatches above the astronauts. These hatches would allow, among other things, for the astronauts to leave their spacecraft in space to perform extra-vehicular activity (EVA), or walks in space.

Pilot control was also designed into the Gemini spacecraft. This delighted the astronauts who didn't always like playing backup to the automatic black boxes that had controlled the flight of the Mercury spacecraft.

FIRST FLIGHTS OF GEMINI

The first flights of the Gemini program were unmanned flights to test the spacecraft and launch vehicle. Everything went as planned, and NASA was ready to launch the first manned flight.

Gemini III Spacecraft Recovery. Astronauts Virgil "Gus" Grissom and John Young were launched in their *Gemini III* spacecraft atop a *Titan II* vehicle on March 23, 1965. After making three Earth orbits in almost five hours of flight, they splashed down in the ocean where Navy helicopters picked them up for transfer to the recovery ship *U.S.S. Intrepid*. Navy frogmen stand on the spacecraft's flotation collar waiting to hook a hoist line to the spacecraft.

On March 23, 1965, astronauts Virgil "Gus" I. Grissom and John W. Young, the first space twins, were launched from Cape Canaveral in the *Gemini III* spacecraft into a three-orbit trip in America's first two-man space flight. They tested the way the spacecraft moved by changing its orbit several times. The flight time of *Gemini III* was four hours, fifty-two minutes, and thirty-one seconds.

WALKING IN SPACE

During the Soviet Union's eighth manned space flight, in March 1965, Lieutenant Colonel Alexei A. Leonov, pushed himself out of an air lock of his *Voskhod II* spacecraft. For ten minutes Leonov maneuvered alone at the end of a sixteen-foot, wire-rope tether. He said, "I felt absolutely free—soaring like a bird." Leonov was the first person to walk in space.

The U.S.'s Turn

Just seventy-seven days later, on June 3, 1965, *Gemini IV* was launched at Cape Canaveral carrying astronauts James A. McDivitt and Edward H. White, II on an epochal 62-orbit flight.

The most exciting part of the *Gemini IV* mission was Edward White's spacewalk. People throughout America watched television spellbound as astronaut White moved out of the *Gemini IV* capsule into space.

White remained outside *Gemini IV* for twenty-one minutes. He was supplied with oxygen through an umbilical cord, which also tethered him to the spacecraft. He was able to maneuver with a Hand-Held Self-Maneuvering Unit that fired small jets of gas. He enjoyed the experience and was reluctant to get back in the spacecraft. Despite the dangers of his adventure, when NASA ordered him to return to his capsule, White said, "It's the saddest moment of my life."

After Soviet pioneer spacewalker Alexei Leonov's ten-minute walk in space in March 1965 and astronaut Ed White's 21-minute cosmic ballet on June 3, 1965, the word "spacewalk" was added to the dictionary. The success of both spacewalks proved that man could work effectively in space.

The *Gemini IV* mission also included extensive maneuvering of the spacecraft by the pilot.

Titan II/Gemini IV Launch.
Titan II missile liftoff from Pad 19 carrying the *Gemini IV* spacecraft with astronauts James McDivitt and Edward White onboard.

Man's First Walk in Space.
On March 18, 1965, cosmonaut Alexei Leonov, aboard the *Voskhod 2* spacecraft, crawled out through an airlock to make the first walk in space. He was outside the spacecraft for ten minutes, tethered by a cable, breathing oxygen from a pack on the back of his spacesuit. For the first time, a man demonstrated that he could perform in orbit independently of his spacecraft. Mongolia issued this stamp to honor the Soviet cosmonaut.

America's First Spacewalker. Astronaut Edward White is seen here performing his spectacular space feat aboard the *Gemini IV* flight. One hundred twenty miles above the surface of the Earth, White slipped out of the spacecraft — the first American astronaut to egress his craft while in orbit. Secured to the *Gemini IV* by a 25-foot umbilical line and a 23-foot tether line, wrapped together with gold tape to form one cord, White remained outside the spacecraft for twenty-one minutes. He held a self-maneuvering propulsion unit in his right hand.

GEMINI V'S EIGHT-DAY FLIGHT

NASA planned longer missions to see how well the astronauts and equipment would perform. Eight days was the minimum duration of a mission to the Moon and back. "Eight days or bust" was the objective — and the crew's motto — when *Gemini V* soared skyward on August 21, 1965 for an eight-day flight. *Gemini V* carried astronauts L. Gordon Cooper, Jr. and Charles Conrad, Jr. This was the first flight to use a new fuel-cell power system. The astronauts performed seventeen science experiments and made five changes in orbit before their capsule safely splashed down eight days later in the Atlantic Ocean. Cooper and Conrad's *Gemini V* orbited Earth a record 120 times, making them the most traveled men in history — although this feat would appear trivial compared with later space efforts. The flight time of *Gemini V* was 190 hours, fifty-five minutes, and fourteen seconds.

Cooper proved that the fine details he had seen on Earth's surface during his Mercury mission (MA-9) had not been a wrong. Both astronauts in *Gemini V* agreed that even wakes of ships at sea and contrails of aircraft could be spotted by eye from orbit. The *Gemini V* crew made an extensive series of photographs of Earth and these findings served as a basis for later orbital surveying of Earth.

After the Gemini VI Splashdown. Safely buoyed by yellow flotation gear attached by U.S. Navy frogmen, *Gemini VI* lies awash fourteen miles from its aircraft carrier recovery ship as astronauts Thomas P. Stafford and Walter M. Schirra climb out of their cramped quarters. The frogman on the far right uses a radiophone to summon the pickup helicopter and the recovery ship.

ORBITING TWINS

The closing days of 1965 saw another remarkable achievement of the Gemini program. The world's first successful rendezvous in space — performed by astronauts Walter M. Schirra, Jr., and Thomas P. Stafford in *Gemini VI*, in conjunction with astronauts Frank Borman and James A. Lovell, Jr., in *Gemini VII* — went on to set the world's duration record for manned orbital flight. Borman and Lovell were launched December 4, 1965, while Schirra and Stafford took off December 15; the latter completed the approach and rendezvous procedures, and landed the following day after sixteen orbits of the Earth. Borman and Lovell remained aloft for another two days, and were recovered December 18. At one point *Gemini VI* and *Gemini VII* were less than one foot apart; they remained close to each other for over five hours. Walter Schirra showed that close-range maneuvering of a spacecraft orbiting at 17,500-miles per hour was, as he described it, "a piece of cake."

In addition to proving that two spacecraft could find one another and dock in space, *Gemini VII* performed twenty science experiments and tested new, lightweight spacesuits.

Gemini V Spacecraft Recovery. U.S. Navy divers exit their helicopter to recover the *Gemini V* spacecraft and astronauts L. Gordon Cooper and Charles Conrad.

Gemini IV Spacecraft Aboard the Recovery Ship. NASA's *Gemini IV* spacecraft undergoes technical observation on the deck of the U.S. Navy aircraft carrier *U.S.S. Wasp* after recovery.

Gemini VII Spacecraft as Seen from Gemini VI. In December 1965 *Gemini VI* and *VII* spacecrafts met in orbit to explore orbital rendezvous techniques, during which the *Gemini VII* spacecraft was photographed through the hatch window of the *Gemini VI* spacecraft. In this first rendezvous in space, the two spacecraft came within one-foot of each other.

Gemini VI Spacecraft in Orbit.
Gemini VI spacecraft in orbit above Earth as seen from the *Gemini VII* spacecraft.

Gemini VII Spacecraft Recovery. Astronauts Frank Borman, command pilot, and James A. Lovell Jr., pilot, sit in a life raft while awaiting pickup by a helicopter from the aircraft carrier *U.S.S. Wasp*. The three-man Navy frogmen team attached the flotation collar to increase the *Gemini VII* spacecraft's buoyancy prior to recovery.

THE FINAL MISSIONS —
DOCKING IN SPACE

The last five Gemini missions accomplished all of the project's objectives: rendezvous, docking, maneuvering, and spacewalking during which work was performed.

Gemini VIII

The first docking of two vehicles in space was accomplished March 18, 1966 by astronauts Neil A. Armstrong and David R. Scott, whose *Gemini VIII* spacecraft was piloted to a rendezvous and docking with a target vehicle. *Gemini VIII* was launched at Cape Canaveral on March 16 and made six and a half orbits, with a flight time of ten hours, forty-one minutes, and twenty-six seconds. After the *Gemini VIII* docked with the Agena rocket, the spacecraft started to spin. During the last orbit, the astronauts steadied the spacecraft, made an emergency reentry, and splashed down safely in the Pacific Ocean. Although both astronauts survived the dangers of space, they suffered from seasickness while waiting for the ship to pick them up.

Gemini VIII after Splashdown. Not long after the *Gemini VIII* spacecraft splashes down a Navy frogmen crew is on hand to attach a flotation collar. Waiting to come out of the spacecraft are astronauts Neil Armstrong and David Scott. The spacecraft came down at an emergency landing point 621 miles south of Yokosuka, Japan.

Agena Target Vehicle. Highlighted by the Sun is the unmanned Agena Target Vehicle that has a docking cone, which was designed to receive the *Gemini VIII* spacecraft's nose. The vehicle has an eight-foot radar antenna.

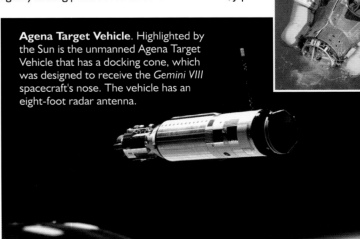

Gemini IX

The next mission continued to practice docking with the Agena rocket. *Gemini IX* was launched June 3, 1966, with astronauts Eugene A. Cernan and Thomas P. Stafford onboard. Astronaut Cernan spent more than two hours outside the spacecraft in extra-vehicular activity. The *Gemini IX* spacecraft made forty-five orbits of Earth and had a flight time of seventy-two hours, twenty minutes, and fifty seconds.

When Cernan and Stafford attempted to dock with a target vehicle, they found the shroud covering the target's nose had failed to jettison and docking was impossible. Unable to dock, the astronauts improvised — using the target vehicle to practice rendezvous maneuvers.

Astronaut Cernan had problems and had to cut short a spacewalk with a new backpack, the astronaut-maneuvering unit (AMU), with a 24-foot long tether. He grew excessively tired and became tangled up in the umbilical line.

Target Vehicle for Gemini IX Mission. The conical shroud of this docking target failed to separate properly, giving it the appearance of an angry alligator, as astronaut Tom Stafford described it. The astronauts had to abandon plans to dock with the vehicle, so astronauts Stafford and Cernan improvised—using the vehicle to practice rendezvous maneuvers.

Gemini IX Astronauts Waiting for the *U.S.S. Wasp*. Astronauts Eugene A. Cernan and Thomas P. Stafford sit with their *Gemini IX* spacecraft hatches open while awaiting the arrival of the U.S. Navy aircraft carrier *U.S.S. Wasp*. *Gemini IX* spacecraft's splashdown was so close to target that the Navy frogmen found the spacecraft still hot to the touch.

Gemini X

Astronauts John W. Young and Michael Collins performed a similar three-day mission in *Gemini X*. It was launched from Cape Canaveral on July 18, 1966. *Gemini X* set a new altitude record of 475 miles during their forty-three orbits of the Earth. The highlight of their mission was Collins' spacewalk by EVA to the Agena rocket, and retrieved a micrometeorite experiment.

Navy Frogman Helping Astronaut. A Navy frogman assists the *Gemini X* astronauts following splashdown at 4:07 p.m., July 21, 1966. Astronaut John W. Young (climbing from spacecraft), command pilot, is the only member of the crew seen in this view.

Gemini XI

The three-day *Gemini XI* mission (September 12-15, 1966), completed by astronauts Charles Conrad, Jr. and Richard F. Gordon, Jr., made the first flight of tethered space vehicles, and reached the highest point of manned space flight in an apogee of 853 miles. Much of the astronauts work was practice for future Moon missions: First rendezvous and docking initial orbit, first multiple docking in space, and extra-vehicular activity (EVA) lasting forty-four minutes by astronaut Gordon. *Gemini XI* made forty-four orbits and had a flight time of seventy-one hours, seventeen minutes, and eight seconds.

"We're over Australia now," command pilot Conrad radioed from *Gemini XI*. "We have the whole southern part of the world at one window. Utterly fantastic."

Going Into Space. The *Gemini XI* spacecraft, atop a Titan II launch vehicle, leaves Pad 19 on September 12, 1966. Astronauts Charles Conrad and Richard Gordon are aboard the spacecraft.

Astronauts and Frogmen Await Pickup. Astronauts Richard F. Gordon, Jr. (left), pilot of the *Gemini XI* space flight, and Charles Conrad Jr., command pilot, sit in life raft while awaiting pickup by a helicopter from the *U.S.S. Guam*. Members of the Navy frogmen team wait with them.

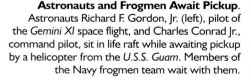

Gemini XII

It was left to astronauts James A. Lovell, Jr. and Edwin E. "Buzz" Aldrin, Jr. to complete the Gemini project with fifty-nine orbits of the Earth. *Gemini XII* was launched November 11, 1966, and had a flight time of ninety-four hours, thirty-four minutes, and thirty-one seconds. During this flight, astronaut Aldrin walked and worked outside of the orbiting spacecraft for more than five and a half hours — a record — proving that a properly equipped and prepared man could function effectively outside of his space vehicle. Unlike astronaut Richard Gordon on *Gemini XI*, Aldrin had no problems with fatigue on his expedition outside the spacecraft. Preflight training conducted underwater had prepared him for the stresses of weightless movement in space. In addition, significant modifications to the hardware — grab rings, handlebars, Velcro patches, and Velcro-covered hand-holds — enabled him to execute his tasks with ease.

The *Gemini XII* flight also demonstrated rendezvous and docking, feasibility of station keeping with a tethered vehicle system, and automatic (computer-controlled) reentry capability. The first photograph of a solar eclipse from space was made on *Gemini XII*.

The flight of *Gemini XII*, in combination with his fourteen days aloft in *Gemini VII*, gave astronaut Lovell a total of more than seven million miles in orbit.

RESULTS OF PROJECT GEMINI

Between March 1965 and November 1966, NASA launched ten successful Gemini flights. During the second manned Gemini flight, Edward White became the first American to perform an extra-vehicular activity (EVA). The next Gemini flight set a record that lasted for five years. The two astronauts aboard the spacecraft orbited Earth for almost eight days. This flight proved that astronauts could live and work in space long enough to go to the Moon and back — a round-trip that would take eight days.

After five Gemini flights, NASA had proven that it could send two astronauts into space for long periods of time. NASA had also shown that astronauts could perform EVAs. The astronauts succeeded in catching and docking with other spacecraft.

Project Gemini had accomplished all of NASA's goals. With ten successful flights, the Gemini program bolstered the nation's confidence in its ability to compete with the Soviet Union in the space war. The Gemini program laid the groundwork for the Apollo program, whose objective was to land humans on the Moon.

America's Two-Man Space Missions. This list shows the Gemini *Titan II* Space Missions at Launch Complex 19 at Cape Canaveral.

Ready to Go! *Gemini XII* astronauts Lovell and Aldrin leave the suiting up trailer prior to their four-day space mission.

Astronauts Lovell and Aldrin Aboard Gemini XII. Astronaut James A. Lovell is photographed inside his spacecraft during the *Gemini XII* mission. Astronaut Aldrin is seen in the background, to the left.

Agena Target Vehicle for Gemini XII. The *Gemini XII* spacecraft, piloted by astronauts Lovell and Aldrin, rendezvous with the Agena Target Vehicle. The 26-foot long Agena was the upper stage of an Atlas Booster that was separately launched.

Gemini XII Splashdown. A Navy frogman from the aircraft carrier *U.S.S. Wasp* leaps from a helicopter into the water to assist in the *Gemini XII* spacecraft recovery operations.

View of a Docked Agena Target Vehicle. Through the open hatch of *Gemini XII* spacecraft, astronaut Edwin Aldrin took this picture showing the docked Agena. The landmass on the horizon is Mexico.

APOLLO PROGRAM

IN 1957, THE SOVIET UNION SUCCESSFULLY LAUNCHED *SPUTNIK 1*, A PRIMITIVE satellite that began the race to the Moon. Less than four years later, the Soviets again managed another spectacular "first" by putting a man into Space. This latest success prompted President John F. Kennedy to challenge the country to be the first to "land a man on the Moon and return him safely to Earth before the end of the decade." This generated the birth of the Apollo program, which eventually overtook the Soviet space program.

Many scientists and engineers at NASA started working on the journey to the Moon. Some were building and testing a rocket that could send a spacecraft all the way to the Moon while others were designing the complex spacecraft that would carry astronauts to the Moon, land them on the lunar surface, and safely return them to Earth. Still others were choosing landing sites on the Moon or developing new equipment for the astronauts.

The astronauts were hard at work, too. They had to be prepared for their incredible journey.

Everyone working on the program recognized that a voyage to the Moon was a tremendous and taxing undertaking that would require hard work, dedication, and tremendous attention to detail. More than 400,000 people worked to accomplish the goal and contributed to tremendous advances in technology.

SATURN V ROCKET

Wernher von Braun's team at the Marshall Space Flight Center in Huntsville, Alabama, was in charge of developing NASA's rockets. In the early 1960s, they began work on a huge multistage, liquid-fuel rocket, — the Saturn V — that would make earlier rockets look like a child's toy.

The enormous Saturn V rocket was far more powerful than the rockets that launched spacecraft and astronauts on the Mercury and Gemini missions. Atop the rockets, three stages rode a three-part spacecraft. The first stage would heave the spacecraft high above the atmosphere, the second stage pushed it into orbit, and the third stage would launch it toward the Moon.

THE APOLLO SPACECRAFT

The Apollo spacecraft consisted of three parts: 1. The Lunar Module that would land on the Moon; 2. The Service Module with its power and support systems; and 3. The Command Module that ferried the astronauts to and from the Moon.

The Service Module (SM) carried fuel, provisions, and the spacecraft's engines. The Command Module (CM) was the astronaut's main travel-

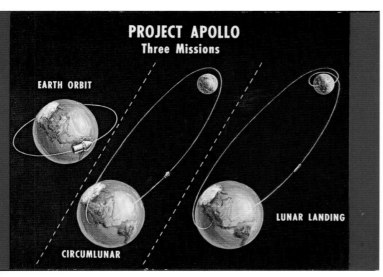

Project Apollo. There were three missions of the Apollo Project. The first one sent astronauts around the Earth to checkout the Command Module; the second, around the Moon to checkout the Lunar Module; and the third, to land men on the Moon.

Apollo 7 Spacecraft Lifts Off Launch Pad. The *Apollo 7* spacecraft, atop a Saturn IB rocket, lifts off from Launch Complex 34. The spacecraft achieved orbit to begin an eleven-day mission. The flight is intended to qualify the Apollo Command Module for a manned flight to the Moon.

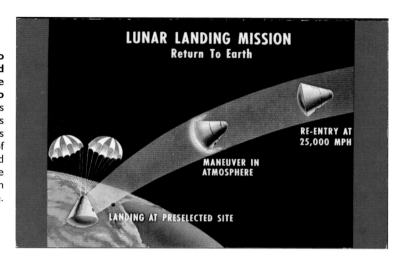

Apollo Command Module Returning to Earth. This view shows an artist's conception of the Command Module returning from Space.

The LM consisted of two sections. On completion of lunar surface operations, the landing legs, rocket engine, and empty fuel tank in the lower descent stage were abandoned to save weight. Using the main fuel tank and engine, only the compact crew module, the ascent stage, lifted off the Moon for the return trip. After making a rendezvous with the Command Service Module (CSM) in lunar orbit, the ascent stage was thrown away altogether, saving yet more weight, and leaving just the CSM for the voyage back to Earth.

APOLLO 1: FIRE IN THE COCKPIT!

Apollo 1 Crew. Crew of the *Apollo 1* spacecraft. Left to right are Edward H. White II, Command Module Pilot; Virgil I. Grissom, Mission Commander; and Roger B. Chaffee, Lunar Module pilot. These astronauts lost their lives in a January 27, 1967 fire in the *Apollo 1* Command Module during testing at the launch facility.

The Apollo project began in tragedy. On January 27, 1967, astronauts Virgil Grissom, Edward White, and Roger Chaffee boarded the spacecraft for a ground test of the equipment. Moments later, a voice came from the spacecraft intercom, "We've got a fire in the cockpit!" Within seconds, the spacecraft was engulfed in flames, feeding voraciously on the pure oxygen atmosphere. On the outside, technicians struggled to open the hatch, but they were too late. The three astronauts couldn't get out before they became asphyxiated. No one could have survived the fire. *Apollo 1* astronauts Grissom, White, and Chaffee were dead.

After the tragic fire, the *Apollo 1* mission was cancelled. No manned space flight would occur for more than one and a half years. Many people doubted that we would get to the Moon on schedule. The nation mourned the horrific death of three of its heroes. After a review of the cause of the fire, NASA ordered extensive modifications of the Apollo Command Module.

Charred Apollo 1 Capsule. On January 27, 1967, tragedy struck the Apollo program. Astronauts Roger Chaffee, Virgil Grissom, and Edward White were taking part in a rehearsal in a command module on the Launch Pad when a sudden flash caused a fire. There was no time to evacuate the astronauts. The atmosphere inside the spacecraft was pure oxygen; the astronauts had no chance of survival.

ing cabin from Earth to the Moon and back. Once the Apollo spacecraft (Command Module, Service Module, and Lunar Module) reached orbit around the Moon, two of the three astronauts would enter the much smaller Lunar Module (LM). Then this mini-spacecraft would separate from the Command and Service Modules (CSM) and descend to the Moon's surface. There, the two-astronaut exploration team would walk on the surface of the Moon. When the Moon visit was over, the LM would use its propulsion to lift-off from the Moon and rejoin the CSM, where they would dock and reenter the CM. They would rejoin the pilot (third astronaut that remained in the CM) and head back to Earth.

APOLLO LUNAR MODULE

Landing on the Moon required a special spacecraft just for that purpose, the Lunar Module (LM). It did not look like a streamlined spaceship from the movies. The LM was the first vehicle designed to operate purely in the vacuum of Space. It was never intended to survive reentry into the Earth's atmosphere. Unlike the main Command Module, it had no need for an aerodynamic shape or heat shielding. This fragile, bug-like machine was created for a very specialized task — landing astronauts on the Moon.

While on the Moon, the LM would be the astronauts' home. For up to three days, they would "camp out" in the LM, sleeping in hammocks hung from the side of the cabin. Once each day the astronauts would put on space suits and backpacks, depressurize the crew compartment, and step outside to walk on the lunar surface.

UNMANNED MISSIONS: APOLLO 2 TO 6

The *Apollo 2* mission consisted of a flight to test the effects of weightlessness on the fuel system.

The *Apollo 3* mission involved a flight to test the rocket and heat shield.

The next three Apollo missions were also unmanned. *Apollo 4, 5,* and *6* involved the testing of the Saturn V rocket and Command and Service Module (CSM), which would carry three-man crews to and from the Moon. The first two missions were very successful; however, *Apollo 6* was a disappointment. From launch, the rocket malfunctioned. The first stage developed a severe oscillation; the second stage lost thrust, and the third stage failed to restart. The mission failed to achieve its prime objective, which was to place its payload in lunar orbit.

APOLLO 7: FIRST MANNED FLIGHT

Apollo 7 was the first manned flight of the Apollo program. The *Apollo 7* spacecraft, atop a Saturn IB rocket, blasted into orbit October 11, 1968. Hope now rested on the redesigned Command Module (CM), which the crew of *Apollo 7* was to test. Astronauts Walter M. Schirra, Jr., Donn F. Eisele, and Walter Cunningham put the spacecraft through its paces and the mission was declared a success. The Command Service Module performed perfectly. The *Apollo 7* spacecraft orbited the Earth 163 times and returned to Earth on October 22, 1968. The mission lasted eleven days.

Saturn V on the Way to its Launch Pad. The 363-foot-tall Apollo Saturn V Space vehicle is leaving the Vehicle Assembly Building (VAB) for Launch Pad 39 at the Kennedy Space Center. The Saturn V stack and its Mobile Launch Tower are atop a crawler transporter.

Apollo 7 Crew.
The crew of the first manned Apollo Space mission, *Apollo 7*, (left to right) are astronauts Donn F. Eisele, Command Module pilot; Walter M. Schirra Jr., Commander; and Walter Cunningham, Lunar Module pilot.

APOLLO 8: OTHER SIDE OF THE MOON

On December 21, 1968, the *Apollo 8* spacecraft, atop the huge Saturn V rocket, roared into Space from its Kennedy Space Center launch pad. This was the first time the Saturn V rocket had been used to carry astronauts into Space. It was also the first spacecraft in history to carry human beings beyond Earth and around another world.

For a frightening few minutes, when astronauts William A. Anders, Frank Borman, and James A. Lovell, Jr., were behind the Moon, they lost contact with NASA. Moments before communications went down, Lovell casually said, "We'll see you on the other side." His message traveled across 240,000 miles of vacant Space

Close-up Apollo 8 Moon View. This is a near vertical photograph of the lunar surface taken with a telephoto lens during the *Apollo 8* lunar orbit mission. The area covered by the photograph is approximately twenty miles on a side; the photographed area is located at about three degrees south latitude and 160 degrees west longitude on the lunar far side.

Apollo 8 Moon View. Nearly full Moon was taken at a point above seventy degrees east longitude. Bright rays radiate from two large craters.

Apollo 8 Crew. A month before their historic trip around the Moon, the *Apollo 8* crew — James Lovell, William A. Anders, and Frank Borman — posed in their spacesuits. They are standing beside the Apollo Mission Simulator at the Kennedy Space Center.

Apollo 8 Liftoff. The first manned Saturn V rocket launched from Launch Complex 39A at the Kennedy Space Center at 7:51 a.m. (EST), December 21, 1968.

Apollo 8 Earthrise. Astronauts Lovell, Anders, and Borman got the sight of their lives as they flew from behind the Moon; they saw the Earth rise above the horizon, 240,000 miles away.

Apollo 9 Mission Commander McDivitt. This close-up view of astronaut James A. McDivitt shows several days' beard growth. The *Apollo 9* Mission Commander was onboard the Lunar Module *Spider* in Earth orbit, near the end of the flight. Astronaut Schweickart took this picture while Scott remained in the Command Module *Gumdrop*. In Earth orbit, the three tested the transposition and docking systems of the Lunar Module and Command Module. On a later mission, astronauts will use what this crew has learned about handling the systems of the two spacecraft.

Apollo 9 Crew. These three astronauts were the crew of the *Apollo 9* mission. They are (left to right) David R. Scott, Command Module pilot; James A. McDivitt, Commander; and Russel L. Schweickart, Lunar Module Pilot.

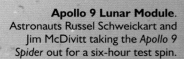

Apollo 9 Lunar Module. Astronauts Russel Schweickart and Jim McDivitt taking the *Apollo 9 Spider* out for a six-hour test spin.

to anxious listeners on Earth. When contact was restored, Lovell's voice came through loud and clear, "Go ahead, Houston, *Apollo 8*." The crew was safe, and they had seen the other side of the Moon.

Below them the Moon looked like wet sand. They saw the crater Tsiolkovsky discovered nine years earlier by the Soviet probe, *Luna 3*. Then, as their spacecraft emerged from the Moon's shadow, they were presented with another site that had never been seen: the Earth rising above the horizon of the Moon. They had also orbited the Moon on Christmas Eve. The *Apollo 8* crew took many photographs of the far side of the Moon, which, unlike the near face's plentiful dark seas, are mostly rocky highlands.

After ten orbits, the spacecraft's engines were fired and the astronauts were on their flight home. The Command Module reentered the Earth's atmosphere, traveling at a speed of 24,791 miles per hour. Communications was established with the recovery vessel *U.S.S. Yorktown*, the parachutes were deployed, and the capsule splashed down in the Pacific Ocean. They were home from the Moon. The date was December 27. By dawn, the recovery team had reached them.

The round-trip to the Moon was a severe test for all Apollo systems and its success gave the program new vigor. Acting NASA administrator Thomas Paine described the mission as "one of the great pioneering efforts of mankind. It is not the end, but the beginning." It proved indeed to be the beginning of the final assault on the Moon that would climax the following year in the first lunar landing.

APOLLO 9: LUNAR MODULE CHECKOUT

The Saturn V rocket blasted off from the Kennedy Space Center March 3, 1969, carrying the *Apollo 9* spacecraft with astronauts James A. McDivitt, David R. Scott, and Russel "Rusty" Schweickart. The astronauts had the task of making sure the Lunar Module (LM) would work in Space.

For this and all subsequent Apollo missions, the crew was allowed to name their craft. The Command Module was dubbed *Gumdrop* and the Lunar Module was named *Spider*.

Russel Schweickart and Jim McDivitt took *Spider* out for a spin, flying for six hours and 111 miles from the Command Module. After the LM voyage, David Scott performed a textbook docking. Russel Schweickart also performed a spacewalk. The mission was a complete success. After orbiting Earth for ten days, one hour, and 151 orbits, the Command Module splashed down in the ocean. The LM had proved itself space-worthy.

APOLLO 10: DRESS REHEARSAL

Apollo 10 Crew. These three astronauts were the crew of the *Apollo 10* mission. From left to right are Eugene A. Cernan, Lunar Module pilot; John W. Young, Command Module pilot; and Thomas P. Stafford, Commander.

Apollo 10 was the dress rehearsal for a lunar landing. On May 18, 1969, the Saturn V rocket boosted the *Apollo 10* spacecraft into Earth's orbit. Aboard the Command Module, named *Charlie Brown*, were astronauts Tom Stafford (Commander), John Young (CM pilot), and Eugene Cernan (LM pilot). Stafford and Cernan took the LM, called *Snoopy*, out for a ride. They descended toward the Moon, but they stayed close enough to the Command Service Module (CSM) so that pilot Young could reach them if they had an emergency.

The astronauts were fascinated by the Moon's surface, describing it as "fantastic," "unbelievable," and "incredible." They took a look at the Sea of Tranquility, where *Apollo 11* would land, and reported that it was smooth with only a few shallow craters. The astronauts had simulated a lunar landing.

Snoopy flew across the Moon's surface for eight hours and ten minutes before re-docking with the Command Module *Charlie Brown*. With Stafford and Cernan safely aboard, Command Module pilot Young turned *Charlie Brown* for home. They left the descent stage of *Snoopy* in lunar orbit and *Snoopy's* ascent stage in Solar orbit, where it remains.

Charlie Brown splashed down on May 26, 1969. On its way home, they transmitted the first color TV pictures sent from Space. They had orbited the Moon thirty-one times.

APOLLO 11 TO 17: MOON LANDINGS

The flights of *Apollo 11* through *Apollo 17* were to be more than astronautical record-setters. They were voyages of discovery. Six missions landed on the Moon and twelve astronauts walked on the lunar surface. They returned with 840 pounds of lunar samples that were analyzed by scientists around the world. The moonwalkers also returned with a wealth of scientific data from experiments performed on the Moon.

FIRST MEN TO THE MOON

I N 1969, THE SATURN V ROCKET BOOSTER LAUNCHED *APOLLO 11* INTO SPACE. The mission was wildly successful and Neil Armstrong was not only the first American, but the first man, to step on the Moon. The succeeding missions, with the exception of *Apollo 13*, gathered increasingly more scientific data until 1972, when *Apollo 17* became the last flight of the Apollo program.

All moonwalks were tightly scheduled. Astronauts had to work quickly, and if they found something unexpected, Mission Control might grant them no more than a few extra minutes before they had to move on.

Astronauts discarded equipment to save weight, but they also left many souvenirs behind. Some, such as *Apollo 11's* gold olive branch of peace, were official NASA items. Others were personal mementos placed on the surface during brief private moments.

The *Apollo 15* crew left a sculpture and plaque in memory of astronauts who had died in the Space race. Charles Duke from *Apollo 16* left behind a photo of his family.

Exploring the Moon was tough and dirty work. At the end of a long day on the lunar surface, astronauts were exhausted and often in pain. They were also dirty, covered with sooty lunar dust. Still, they were all sad to have to leave.

The last part of the Moon mission is a dangerous one. The Command Module hits Earth's atmosphere at 24,500 miles per hour — ten times faster than a rifle bullet. If it hits at the wrong angle, it will either burn up or shoot off into Space, never to return. If the heat shield fails, the intense heat of reentry will be fatal. If the parachutes fail to open, the capsule will hit the ocean too hard. If they land too far off course,

Earth Rise.
Swathed in white clouds, Earth rises over the lunar horizon in this dramatic view from *Apollo 11*, in orbit around the Moon. The Moon orbits Earth at the rate of 50,800 miles per day, even as Earth covers 1.5 million miles a day in its travels around the Sun.

Next Stop — Launch Pad 39A.
Apollo 11 Commander Neil A. Armstrong leads astronauts Michael Collins and Edwin A. Aldrin, Jr. from the Manned Spacecraft Operations Building to the transfer van for the eight-mile trip to Launch Pad 39A.

Apollo 11 Crew. The three-man crew of the *Apollo 11* lunar landing mission was (left to right) Neil A Armstrong, Commander; Michael Collins, Command Module pilot; and Edwin E. Aldrin, Jr., Lunar Module pilot.

Far Side of the Moon. An oblique view of the lunar far side, looking southwest, photographed from the *Apollo 11* spacecraft in lunar orbit. The large crater in the center is International Astronomical Union #308. This crater has a diameter of approximately fifty statue miles.

they could sink before the rescue ship reaches them. However, for all of the Apollo missions, the splashdowns went as planned.

Eight minutes after first entering the atmosphere, the Command Module slows down enough for three large parachutes to open. At splashdown, balloons inflate to keep the capsule upright. Navy divers help the astronauts get into life rafts or baskets to be hoisted into a helicopter that takes them to the waiting rescue ship.

APOLLO 11

The *Apollo 11* mission began at 1:32 p.m. on July 16, 1969, with astronauts Neil A. Armstrong, Edwin "Buzz" Aldrin, and Michael Collins perched on top of a 363-foot Saturn V rocket as 7.5 million pounds of thrust blasted them into Space. Once in Earth orbit, the third and final stage of the Saturn V shut down. *Apollo 11* then swung around the Earth and the third stage reignited to propel the craft on its journey to the Moon. The spacecraft entered lunar orbit on July 19. It made thirty orbits, passing over the planned landing site in the Sea of Tranquility that had been selected by the *Apollo 10, Ranger,* and *Surveyor* spacecraft.

It was a time of high human drama and incredible technological achievement — a time when man stood on the brink of a new frontier as all the world looked on in awe. On July 20, 1969, astronauts Armstrong and Aldrin piloted their bug-like spacecraft, named *Eagle*, down toward the surface of the Moon and a rendezvous with destiny.

People of many nations seemed transfixed as they witnessed, via radio and television reports, the last dramatic moments of the Lunar Module (LM) descent to the lunar landscape. They knew history had been made when Armstrong reported across 240,000 miles of Space, "The Eagle has landed." Hours later, Armstrong spoke to the world as he placed his foot on the surface of the Moon, saying, "It's one small step for man, one giant leap for mankind."

At first, Armstrong was tethered to the ladder because no one knew if the surface would be like quicksand, but he soon found that it was easy to walk around. Armstrong's immediate task was to grab a sample of rocks and soil, in case he had to leave in a hurry. All went perfectly. Aldrin joined Armstrong on the surface about twenty minutes later and helped to set up experiments and gather more rocks.

The astronauts set up three science experiments on the Moon: a reflector to bounce laser beams back to Earth, a foil "flag" to collect solar particles, and a seismometer to measure moonquakes. They also deployed a television camera so that it could take panoramic views of the Moon.

Walking on the Moon. *Apollo 11* astronaut Buzz Aldrin walking near the Lunar Module *Eagle* on the surface of the Moon. Reflected in his helmet are the *Eagle* and astronaut Neil A. Armstrong.

Liftoff of Apollo 11 Spacecraft. Man's historic first journey to the surface of the Moon begins atop a pillar of flame as *Apollo 11* lifts off Launch Pad 39A at the Kennedy Space Center on July 16, 1969.

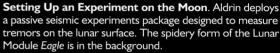

Eagle and Columbia Linkup in Lunar Orbit. As the Earth rises above the Moon's horizon, the Lunar Module *Eagle* approaches the Command Module *Columbia* from which astronauts Michael Collins took this photo. The Lunar Module's descent stage, which was used as a launch platform, was left on the lunar surface. After the *Eagle* and *Columbia* linked up, Armstrong and Aldrin rejoined Collins and prepared for the return trip to Earth.

Aldrin used the geologist's hammer to take soil samples and Armstrong photographed the Lunar Module so that engineers could use the photographs to assess how it had dealt with the rigors of landing. They planted the American Flag and took an unscheduled phone call from President Richard Nixon. They also left memorabilia, including an Apollo mission badge, an information disk, Soviet metals, and a plaque bearing the inscription, "Here Men from the Planet Earth First Set Foot Upon the Moon July 1969 A.D. We Came in Peace For All Mankind."

Second Man to Walk on the Moon. *Apollo 11* Astronaut Edwin E. (Buzz) Aldrin, Jr., climbs down from the Lunar Module for his walk on the Moon. Neil A. Armstrong, the first astronaut to set foot on the lunar surface, took the photo.

The First American Flag on the Moon. Aldrin poses beside the deployed American flag on the Moon. The Lunar Module *Eagle* is on the left. The astronauts' footprints in the soil of the Moon are clearly visible in the foreground. Armstrong used a 70-mm lunar surface camera in taking this picture.

Famous Footprint. Aldrin photographed his footprint to record the properties of lunar soil. His boot left a sharp outline in what looked like fine talcum powder. The image has become one of the most famous photos ever taken — a symbol of the human need to explore. Man's footprints will rest forever on the Moon's surface. In 1969, the U.S. Postal Service used the image on a 33-cent stamp.

Setting Up an Experiment on the Moon. Aldrin deploys a passive seismic experiments package designed to measure tremors on the lunar surface. The spidery form of the Lunar Module *Eagle* is in the background.

Watching the Apollo 11 Launch. Former President Lyndon Johnson and Vice President Spiro Agnew view the liftoff of the *Apollo 11* from the stands at the Kennedy Space Center (KSC) VIP viewing site. The two political figures were at KSC to witness the launch of the first Manned Lunar Landing mission.

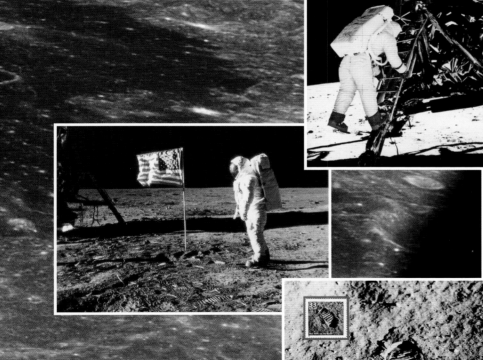

Apollo 11 Splashdown. When the *Columbia*, carrying the three astronauts, splashed down southwest of Hawaii on July 24, it marked the fulfillment of President John F. Kennedy's challenge to send a man to the Moon and return him safely to Earth before the end of the decade. *Apollo 11* splashed down at 11:49 a.m. on July 24, 1969 — about 812 nautical miles southwest of Hawaii and twenty-four miles from the recovery ship.

Back on Earth. The three *Apollo 11* astronauts, safely home after splashdown in the Pacific Ocean, get ready to enter a life raft for transfer to the U.S. Navy aircraft carrier the *U.S.S. Hornet*.

Welcome Home! President Richard M. Nixon was in the central Pacific recovery area to welcome the *Apollo 11* astronauts aboard the *U.S.S. Hornet* recovery ship for the historic lunar landing mission. Already confined to the Mobile Quarantine Facility (MQF) are (left to right) Armstrong, Collins, and Aldrin. The three lunar travelers remained in the MQF until they arrived at the Manned Spacecraft Center's Lunar Receiving Laboratory.

After spending two and a half hours on the lunar surface, Armstrong and Aldrin reentered the Lunar Module *Eagle*, discarded some unnecessary items, and began the trip back into lunar orbit and a rendezvous with the Command Module, named *Columbia*, and astronaut Collins. They docked with *Columbia*, jettisoned *Eagle* and, carrying 47.2 pounds of lunar rock, began the journey home.

Splashdown took place in the Pacific Ocean, twenty-four miles away from the recovery ship *U.S.S. Hornet*. The crew were immediately taken into a quarantine facility, through the armored window of which they talked again to President Nixon, who joined the *Hornet* to meet them. They were home, brought "safely back to Earth."

APOLLO 12 TO 14

Every mission after *Apollo 11* became more ambitious, landing at more mountainous sites, with astronauts traveling further from the Lunar Module. On November 19, 1969, *Apollo 12's* Lunar Lander *Intrepid* landed on the Ocean of Storms, near the *Surveyor 3* Space probe. The astronauts visited *Surveyor 3*, which was coated with dust.

In April 1970, the *Apollo 13* mission was cancelled after an explosion, but later that year, *Apollo 14's* landing craft, *Antares*, touched down near the crater Fra Mauro. To the amusement of TV viewers on Earth, astronaut Al Shepard whacked a golf ball with a soil sampling stick. *Apollo 14* astronauts also wheeled a tool cart to collect debris from a young impact crater.

Apollo 12

At 11:32 a.m. (EST) on November 14, 1969, *Apollo 12* took off in a storm — thirty-six and a half seconds after launch, the Saturn V rocket was hit by lightning. The mission was seconds away from abort. After flicking a few switches, the astronauts corrected the problem. Aboard the *Apollo 12* were astronauts Charles Conrad, Richard Gordon, and Alan Bean. On this gray November morning, President Nixon took his seat in the stands to watch the launch. *Apollo 12* would be the only lunar mission launch cheered on by a serving president.

On November 19 the Lunar Module *Intrepid*, with Conrad and Bean aboard, unlocked from the Command Module *Yankee Clipper* and began the descent to the Moon. Bean and

Apollo 12 Crew. These three astronauts were the crew of *Apollo 12*. From left to right are Charles Conrad, Jr., Richard F. Gordon, Jr., and Alan L. Bean.

President Nixon Greets Apollo 12 Launch Watchers. President Richard M. Nixon greets the crowd that has gathered to watch the *Apollo 12* launch at Kennedy Space Center.

Conrad's time on the Moon was largely spent performing two walks of the Moon, deploying equipment, retrieving parts of *Surveyor 3* for examination on Earth, and making jokes.

After spending thirty-one hours and thirty-one minutes on the Moon, they were back in the Lunar Module and on the way back to the Command Module *Yankee Clipper* to rejoin astronaut Richard Gordon. They jettisoned the Lunar Module, which crash-landed on the Moon. Splashdown occurred on November 24. The astronauts brought back the *Surveyor 3's* defunct camera, seventy-six pounds of lunar samples, and enough traveler's tales for a lifetime.

Apollo 13

Astronauts James Lovell, John Swigert, and Fred Haise, the crew of *Apollo 13*, blasted off into Space on April 11, 1970. They did not know that their Command and Service Module, *Odyssey*, had a problem that had been with them since before takeoff.

On their way to the Moon on April 13, the *Apollo 13* astronauts had just finished a television broadcast to Earth when they heard a loud bang. The meters for oxygen and power levels went to zero. To survive the trip home, the three astronauts had to leave the crippled Command Module, now losing oxygen and power, and live for four days in the Lunar Module,

Apollo 13 Crew. These three astronauts were the crew of *Apollo 13*, the third lunar landing mission. From left to right are James A. Lovell, Jr., John L. Swigert, Jr., and Fred W. Haise, Jr.

a spacecraft designed to house two astronauts for just two days. The astronauts fired the Lunar Module's engine (designed for landing on the Moon) to put them on a path that looped around the Moon and then returned directly to Earth. They jettisoned the Lunar and Service Modules, powered up the Command Module on its batteries, and reentered Earth's atmosphere. On April 17, four days after the crisis began, the *Apollo 13* Command Module plunged back through the Earth's atmosphere, opened its orange striped parachutes, and dropped into the Pacific Ocean just three and a half miles from the *U.S.S. Iwo Jima* rescue ship, bringing the astronauts safely home.

As an aborted mission, *Apollo 13* was officially classified as a failure. However, it was certainly the most brilliant demonstration of human capability under stress of all the Apollo flights.

This incident was also the basis of the popular 1995 movie, "Apollo 13," starring Tom Hanks, and origin of the phrase, "Houston, we have a problem." Though the exact phrase said by astronaut Lovell was, "Houston, we've had a problem," the oft-quoted tagline of the movie became "Houston, we have a problem."

Apollo 14

Apollo 14 blasted off from Kennedy Space Center on January 31, 1971, carrying astronauts Al Shepard, Stu Roosa, and Ed Mitchell. They were bound for Fra Mauro, a cratered highland area that geologists hoped would provide samples of the earliest bedrock of the Moon.

The astronauts stayed longer, conducted more experiments, and tested the endurance of their spacesuits and themselves more than anyone had done before.

Apollo 14 Crew. These three astronauts are the crew of the *Apollo 14* lunar landing mission. They are Alan B. Shepard, Jr. (center), Stuart A. Roosa (left), and Edgar D. Mitchell. The *Apollo 14* emblem is in the background.

They also sent the first color TV transmission from the Moon's surface. The return to Earth on February 9th was uneventful and the splashdown was close to the planned point. They brought back samples from the crater Fra Mauro, and Shepard got to play golf on the Moon's surface using his golf ball and an improvised club.

Astronaut Activity on the Moon. Astronaut James B. Irwin, saluting the flag at the Hadley-Apennine landing site on the Moon. In the center is the Lunar Module *Falcon*. On the right is the Lunar Rover. This remarkable Moon buggy allowed the astronauts to explore areas several miles away from their landing site.

APOLLO 15

The all-U.S. Air Force crew of *Apollo 15* — astronauts Dave Scott, Al Worden, and Jim Irwin — were lifted into Space July 26, 1971. They carried the heaviest payload, had the most focused agenda, and would stay longer and travel further than any of the previous Apollo missions. The mission-landing site was near to Mount Hadley, where they would be able to investigate terrain that was thought to have been formed by volcanic activity.

On July 30 the Lunar Module *Falcon* began its descent to the surface of the Moon. Above them, in Command Module *Endeavor*, astronaut Worden would be busy taking photographic examination of the ground beneath his orbiting spacecraft. The *Falcon*, containing astronauts Scott and Irwin, shortly thereafter deployed the Lunar Rover — Scott and Irwin became the first men to drive on the Moon's surface.

The Lunar Rover, which looked a little like a dune buggy and sometimes called a Moon buggy, allowed astronauts to travel up to five miles from the Lunar Module, far enough to visit interesting craters, but near enough to walk back to the lander if the Rover broke down (although it never did). Each wheel had its own battery-powered electric motor, providing four-wheel drive and four-wheel steering. A television camera and dish antenna sent images back to Earth from each stop along the way.

Scott and Irwin explored the shoreline of the Sea of Rains. They visited a winding valley called the Hadley Rille and found a patch of green soil that had been sprayed out of a lava fountain. They showed that objects of different weights fall at the same speed on the airless Moon. They dropped a hammer and a falcon's feather from the same height and both hit the Moon's surface at the same time. They used the Lunar Rover to drive up the side of towering Mount Hadley.

The *Falcon* blasted off from the lunar surface on August 2nd. This had been the most productive lunar mission to date, and was hailed as "man's greatest hours in the field of exploration." The astronauts brought back 171 pounds of lunar samples. Splashdown took place on August 7, 1971. The *Apollo 15* mission had lasted twelve days, seven hours, and eleven minutes.

Apollo 15 Crew. Posing with the Lunar Rover Vehicle are, from left to right, astronauts James B. Irwin, David R. Scott, and Alfred M. Worden. The amazing electric Lunar Rover was used on the final three Apollo missions. This vehicle, powered by two 36.5-volt batteries, with tires of woven piano wire faced with titanium chevrons, was able to ride on the lunar surface without sinking in the deep layer of dust present at the site. It also contained a gyroscopic navigational system and a radio unit that allowed the astronauts to talk directly to Earth.

APOLLO 16

With astronauts John Young, Ken Mattingly, and Charlie Duke, *Apollo 16* made a perfect lift-off from the Kennedy Space Center on April 16, 1972. Just under four days later, the Lunar Module *Orion* landed as planned near Crater Descartes. Young and Duke curbed their impatience and slept before their hard work began.

Fully rested, they unpacked and assembled their Lunar Roving Vehicle and began the first of three long EVAs. The astronauts visited a hilly region called the Descartes Highlands and, using the Lunar Rover, they explored Stone Mountain and North Ray crater. The crater's inside walls were found to be layered, where ancient lava flows had built up over time.

With the help of their Lunar Rover, Young and Duke had traveled almost seventeen miles across the lunar surface in three separate excursions.

By the time Young, Duke, and Mattingly returned to Earth, their flight duration of 265 hours, fifty-one minutes, and five seconds made them the longest Apollo mission so far, and included a record twenty hours and fourteen minutes of moon walking. Their 207 pounds of rock samples set another record.

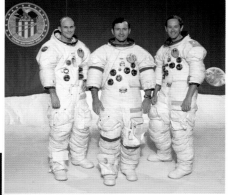

Apollo 16 Crew. These three astronauts were the crew of the *Apollo 16* lunar landing mission. They are, left to right, Thomas K. Mattingly II, John W. Young, and Charles M. Duke, Jr.

Vehicles on the Moon. The Lunar Rover, parked near the Lunar Module *Orion*, is ready to explore the lunar surface with astronauts Young and Duke.

APOLLO 17

This mission was the last in the Apollo series. While Eugene Cernan and Ronald E. Evans were military men, Dr. Harrison "Jack" Schmitt was a trained geologist. These astronauts aboard *Apollo 17* went on the only mission in which the Saturn V rocket was launched at night. Its trail could be seen seven hundred miles away and was so bright that it brought fish off Cape Cod to the surface, fooled into thinking it was the Sun. The night lift-off occurred on December 7, 1972.

The destination was the Taurus-Littrow area of the Moon. A Moon landing was made on December 11. They were a mere 330 yards short of the target and had two minutes of fuel left in the tank. Cernan and Schmitt spent three days on the Moon and made three seven-hour excursions.

They visited an area near the crater Littrow and discovered some bright orange soil, which was later found to be made of tiny colored beads formed in the intense heat of a meteorite impact. Astronauts drove nearly five miles from the Lunar Module in their longest treks. They collected samples at the foot of the Taurus Mountains, but it wasn't all work and no play — Schmitt and Cernan hopped, skipped, and joked their way around the Moon.

The astronauts left behind a plaque bearing a solemn message:

"Here man completed his first explorations of the Moon, December 1972 A.D. May the spirit of peace in which we came be reflected in the lives of all mankind."

After docking with the Command Module *America*, the crew of three orbited the Moon for two days and then headed for Earth. Splashdown occurred December 19; the crew was picked up by the aircraft carrier *U.S.S. Ticonderoga*. The duration of the *Apollo 17* mission was twelve days, thirteen hours, and fifty-one minutes.

The crew brought back a total of 243.65 pounds of Moon rock, among which was the Goodwill Rock that was to be sliced up and given to the students of seventy countries.

Apollo 17 Crew. The *Apollo 17* crew poses with the Lunar Roving Vehicle. Astronaut Eugene A. Cernan sits at the controls, Dr. Harrison "Jack" Schmitt (left), and Donald A. Evans. In the background is the Saturn V booster that will send them to the Moon.

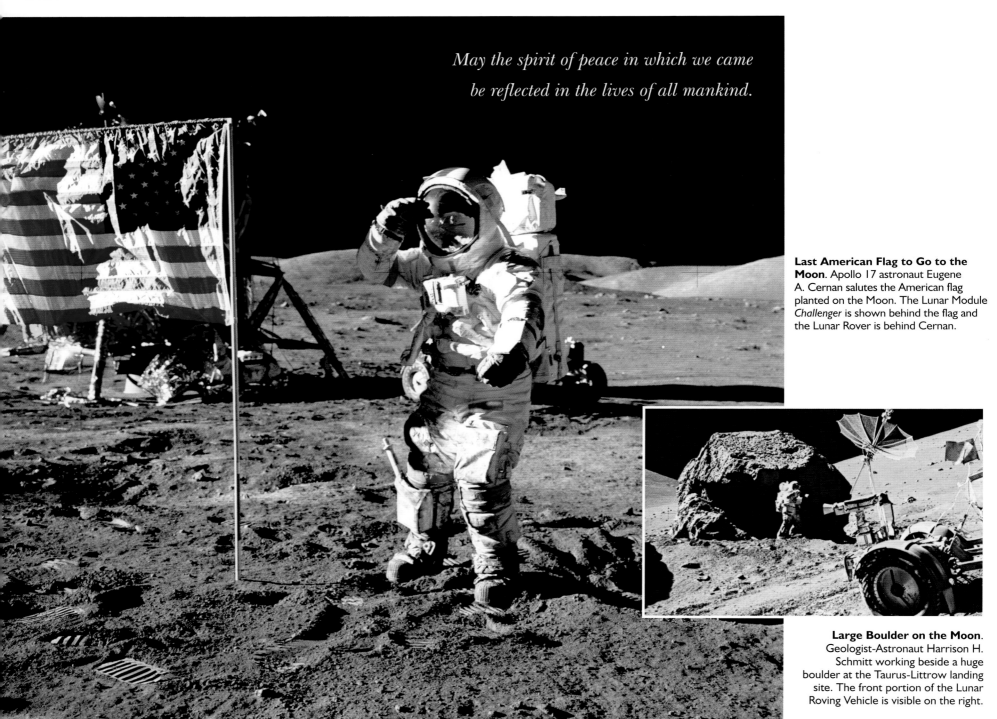

May the spirit of peace in which we came be reflected in the lives of all mankind.

Last American Flag to Go to the Moon. Apollo 17 astronaut Eugene A. Cernan salutes the American flag planted on the Moon. The Lunar Module *Challenger* is shown behind the flag and the Lunar Rover is behind Cernan.

Large Boulder on the Moon. Geologist-Astronaut Harrison H. Schmitt working beside a huge boulder at the Taurus-Littrow landing site. The front portion of the Lunar Roving Vehicle is visible on the right.

SKYLAB: AMERICA'S ORBITAL SPACE STATION

SKYLAB WAS AMERICA'S FIRST MANNED EXPERIMENTAL SPACE STATION AND, at the time, the largest spacecraft ever placed in Earth orbit. It was launched May 14, 1973, at the Kennedy Space Center atop a Saturn V rocket. The major goal of *Skylab* was to determine if people could physically withstand extended stays in space and continue to work.

Other mission objectives were to evaluate techniques designed to gather information on the Earth's resources and to conduct a major investigation of the Sun using the Space Station's special solar telescopes. Experiments were to be carried out in four major areas: solar physics, life sciences, Earth observations, and material science.

SALVAGED FROM THE APOLLO PROGRAM

Skylab was built from the upper stage of the massive *Saturn V* Moon rocket. The Saturn's hydrogen tank was converted into a spacious two-story accommodation for a three-man crew. *Skylab* was the first space vehicle to offer creature comforts.

The body dimensions of *Skylab* was eleven by twenty-seven feet, weighed one hundred tons on Earth, and had 12,000 cubic feet of volume — about the size of a small house. The largest section was the 10,246-cubic foot workshop, called the Orbital Work Shop (OWS), which provided crew quarters, experiment compartment, and storage for most of the expendables such as water and food. The workshop had 20,000 or so pieces of equipment. *Skylab* had facilities for heating meals, five freezers, water for showers and other needs, and lockers that stored towels, soap, and changes of clothing. *Skylab* was launched with a fixed amount of supplies — food, clothing, water, and air. Astronauts slept in zipped up sleeping bags. (In weightlessness, beds are unnecessary.) They ate a wide variety of foods, some in cans and some in bags, which mixed with hot water to make nourishing soups and stews. Electrical power was provided by solar panels.

In front of *Skylab* was an airlock module and the docking adapter, to be used by an Apollo CSM (Command Service Module) spacecraft. The airlock was the environmental, electrical, and communications control center. It held the nitrogen and oxygen tanks that supplied the atmosphere in Skylab, which was maintained at about a third of the atmospheric pressure on Earth.

Skylab Space Station Launch. At noon on May 14, 1973 the last *Saturn V* rocket to be launched lifts off Launch Complex 39A, bound, not for the Moon this time, but for Earth orbit. It carried the *Skylab Space Station*, America's first orbital Space Station.

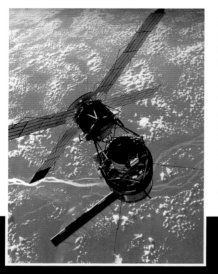

Skylab Space Station. This photograph depicts the *Skylab Space Station* as seen from the Apollo spacecraft, just prior to docking. Weighing more than one hundred tons (on Earth), it stretched 118 feet from end to end. It was as large as a small house and had most of the same comforts. The three passengers had to share a shower and lavatory, but each had a private compartment for sleeping. Also aboard were medical and scientific equipment, 720 gallons of drinking water, and more than 2,000 tons of food. The purpose of *Skylab* was to see if man could live and work in space for the long periods of time it would take to travel back and forth to other planets and moons.

Overhead View of Skylab. The *Skylab Space Station* in Earth orbit as photographed from the third visiting *Apollo* spacecraft. During launch on May 14, the micrometeoroid shield on the Space Station ripped away and caused the loss of one of the solar arrays. The gold parasol visible was designed to protect the Space Station against solar heating.

SKYLAB CREWS

Skylab circled the world unoccupied for almost two weeks until a crew of three followed in an Apollo capsule, launched from a Saturn 1B. The first crews arrived eleven days after *Skylab* went into Earth orbit. Three crews visited the Space Station with missions lasting twenty-eight, fifty-nine, and eighty-four days (a record not broken by an American astronaut until the Shuttle-Mir program over twenty years later). The astronauts performed X-ray studies of the Sun, remote sensing of the Earth, UV astronomy experiments, and medical and biological studies.

SKYLAB MISSIONS

The *Skylab Space Station* was launched into a near-circular orbit 270 miles above the Earth's surface. *Skylab* came very close to failure on its first day in Space. During the launch, a micro-meteoroid shield tore loose and damaged the main solar array that was to supply the station's electricity. Even more potentially damaging — the loss of the micro-meteoroid shield exposed the station to soaring temperatures that threatened to destroy the $200 million space station.

Using Apollo spacecraft and Saturn 1B rockets, three separate launches sent three-man crews to occupy *Skylab*.

Skylab in Orbit. The *Skylab Space Station* in Earth orbit taken by one of the astronauts aboard the second visiting *Apollo* spacecraft.

Skylab Mission Patches.

Skylab
Date: May 14, 1973
Astronauts: Unmanned

Official emblem for the NASA Skylab program. The emblem depicts the United States *Skylab Space Station* cluster in Earth orbit with the Sun in the background. The sun was one of its main research objects.

First Crew to Skylab
Date: May 25-June 22, 1973
Astronauts: Charles Conrad, Joe Kerwin, Paul Weitz
Mission Duration: 28:00:49

First *Skylab* crew mission patch.

Second Crew to Skylab
Date: July 28-September 25, 1973
Astronauts: Alan Bean, Owen Garriott, Jack Lousma
Mission Duration: 59:11:09

Second *Skylab* crew mission patch.

Final Crew to Skylab
Date: November 16, 1973-February 8, 1974
Astronauts: Gerald Carr, Ed Gibson, Bill Pogue
Mission Duration: 84:01:16

Final *Skylab* crew mission patch.

First Skylab Crew

The first Skylab crew was blasted into space by a Saturn 1B rocket at 9:02 a.m. on May 25, 1973. Aboard the spacecraft was mission commander Pete Conrad, science pilot Joseph Kerwin, and pilot Paul Weitz. The first Skylab crew was known as the "Fix-It Crew" because they were required to fix many of the hardware problems that occurred during *Skylab's* launch. Astronaut Conrad boasted, "We can fix anything," and the crew made good on the boast by fixing many of the problems on the crippled station, including installing a parasol and freeing a jammed solar wing.

First Skylab Apollo/Saturn 1B Launch. The first Skylab crew leaves Complex 39B at 9:02 a.m., May 25, 1973. Astronauts Conrad, Kerwin, and Weitz became the first crew to work and live in the orbiting *Skylab Space Station*.

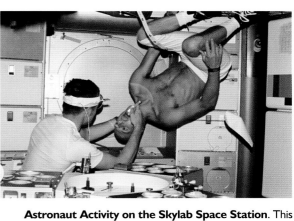

Astronaut Activity on the Skylab Space Station. This photograph shows astronaut activity under zero gravity conditions aboard the *Skylab Space Station*. Scientist/astronaut Joseph P. Kerwin, science pilot and a doctor of medicine, gives an oral physical examination to astronaut Charles Conrad, Jr., commander. Conrad almost literally stands on his head in the weightlessness of space with only a restraint around his left leg holding him in position. Note the floating piece of paper at the right. The photograph was taken by astronaut Paul J. Weitz, pilot, the third member of the first *Skylab* crew.

First Skylab Apollo Spacecraft Leaving the VAB. This view shows the first *Skylab Apollo* spacecraft atop a Saturn 1B rocket, leaving the Vehicle Assembly Building (VAB, on its way to Launch Complex 39B.

The astronauts started up the EREP (Earth Resources Experiment Package), which contained cameras and other instruments that could section and survey a continent while distinguishing a single house or cornfield, and activated the eight telescopes in the twelve-ton solar observatory.

The crew rode an exercise bike thirty minutes a day and took tests to gauge the disorientation of weightlessness. They took turns working, slept in sleeping bags in their own closet-sized bedrooms, used an air-suction toilet, and took weekly showers (three quarts of water per shower).

After spending twenty-eight days and fifty minutes in space, the "Fix-It Crew" came back changed men. All three lost weight; their hearts had shrunk about three percent and raced when required to pump on Earth again. However, two days after splashdown all had recovered.

Second Skylab Crew

On July 28, 1973, one month and six days after the first crew splashed down, the second *Skylab* team blasted off. Their projected stay was fifty-nine days, more than twice that of first *Skylab* crew.

The second crew consisted of U.S. Navy Captain Alan Bean (Pete Conrad's pilot on Apollo 12), electrical engineer and physicist Owen Garriott, and U.S. Marine pilot Jack Lousma. Both Garriott and Lousma were rookies, but Lousma was so calm he dozed during the countdown.

The crew experienced space sickness for about four days. They exercised twice as much as the first crew, completed thirty-nine Earth surveys, logged 305 hours at the solar telescopes, conducted nineteen experiments, and traveled twenty-four and a half million miles. The crew splashed down on September 25; they had spent fifty-nine days and eleven hours in orbit.

Astronaut at the Telescope Control Console. Astronaut Owen K. Garriott, science pilot of the second *Skylab* crew, sitting at the control console of the solar physics telescopes.

Final Skylab Crew

At 9:01 a.m. on November 16, 1973, an all-rookie crew lifted off for their twelve-week sojourn in Space. They were U.S. Marine Lieutenant Colonel Gerald Carr (commander); U.S. Air Force Lieutenant Colonel William Pogue (pilot); and engineer and physicist Edward Gibson (science pilot).

The third and final Skylab crew made many equipment repairs in space. This ability of the astronauts to isolate and correct faults as they occurred added considerably to the overall success of the Space Station mission.

The astronauts watched the seasons change in two hemispheres, as wheat ripened in the Argentine summer and northern rivers grew swollen with ice. They photographed a smoking volcano in Japan, and at night, saw the lights of the Eastern United States.

One astronaut, who spent long hours at the solar observatory, witnessed and filmed the shooting orange lava of a solar flare — the first one to be recorded from beginning to end. A solar flare results from gigantic energies released in the destruction of magnetic fields that are above the solar surface.

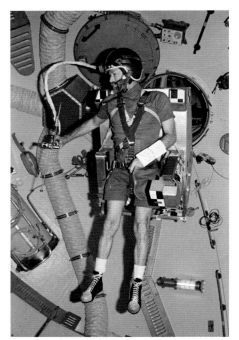

On Christmas Day, astronauts Pogue and Carr took a record seven-hour spacewalk during which they observed and photographed the bright comet Kohoutek as it rounded the Sun. This was a scientific bonus for these photographers.

During the crew's voyage in space they made high-quality alloys and crystals in the Space Stations electric furnace.

The last *Skylab* crew returned to Earth February 8, 1974 and was picked up by the *U.S.S. New Orleans*. The *Skylab Space Station* was left in a parking orbit.

Maneuvering Equipment Experiment. Astronaut Gerald P. Carr, commander of the final *Skylab* crew, conducting a maneuvering experiment aboard the *Skylab Space Station*.

BENEFITS OF SKYLAB PROJECT

After the final crew left on February 8, 1974, the station was put into a stable altitude and shut down. It was expected to remain in orbit for eight to ten years, but increased atmospheric draw caused it to reenter the atmosphere on July 11, 1979. The debris dispersion area covered the southeast Indian Ocean and a sparsely populated area of Western Australia.

The end of the Skylab Project was just the beginning for scientists down on Earth, as they would be evaluating the data from the Space Station for years. More than 180,000 photographs were taken of the Sun. Among other discoveries, they revealed a star much more complex than had been thought. Earth surveys uncovered new deposits of oil, ore and water. Over 40,000 photographs were used to improve weather predictions, crop forecasting, and water and forestry planning.

Skylab carried the largest collection of scientific hardware ever flown in Space. Astronauts carried out more than three hundred scientific investigations in virtually every field that could benefit from the unique vantage point *Skylab* offered.

The Skylab Project also determined that astronauts could grow two inches taller while in space because of weightlessness. Because an astronaut's spine does not have to support the body's weight in space, it can stretch slightly. After returning to Earth, the astronaut's spine shrinks to its original length as it again supports the body's weight.

Perhaps the most significant knowledge gathered from *Skylab* was that people can survive in space without physiological problems and that some industrial processes have their greatest potential in space. Overall, the Skylab program played a major part in the transition from the Apollo era of the 1960s to the Space Shuttle era of the 1980s and beyond.

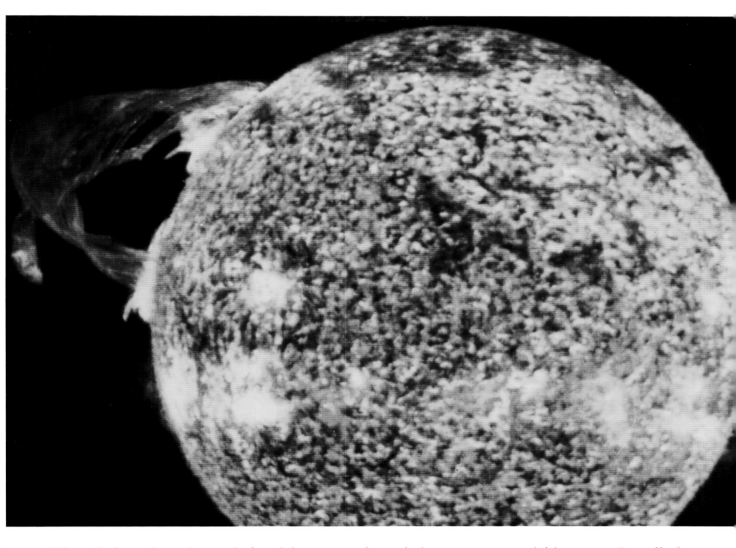

The Sun. Photographed by the final *Skylab* crew, at the upper left can be seen one of the most spectacular solar flares ever recorded, spanning more than 367,000 miles. A photograph taken seventeen hours earlier showed this feature as a large, quiescent mass on the Sun's eastern side. The solar poles are distinguished here by their dark tone and relative absence of super-granulation network.

THE APOLLO-SOYUZ TEST PROJECT

T HE APOLLO-SOYUZ TEST PROJECT (ASTP) WAS HISTORY'S FIRST international manned spaceflight. It was the result of six years of joint decisions and negotiations that took place in both the United States and the Soviet Union.

The scientific experiments conducted during the Apollo-Soyuz mission enhanced the prospects for future international space cooperation. The historic mission marked the way for future joint manned flights.

Soviet Soyuz Spacecraft. A view of the *Soyuz* spacecraft in Earth orbit, photographed from the Apollo spacecraft during the joint U.S./U.S.S.R. Apollo-Soyuz Test Project docking in Earth orbit mission. The *Soyuz* is contrasted against a white-cloud background in this overhead view. The *Soyuz* spacecraft had a two-man crew. Alexei Leonov, who had pioneered spacewalking in orbit from a *Voshkod 2* space vehicle in March 1965, was the commander. Traveling with him was Valeriy Kubacov, who had previously flown on the *Soyuz 6* mission in 1969.

MEETING IN SPACE

At 8:20 a.m. (Eastern Daylight Time) on July 15, 1975, the U.S.S.R. *Soyuz* spacecraft, carrying two cosmonauts, thundered into orbit from its launch base at Kazakhstan. Over seven hours later, at the Kennedy Space Center, the launch control center for Launch Complex 39B, fired the Apollo spacecraft on its final mission. After two days of orbital

Apollo-Soyuz Saturn 1B Launch. NASA used a Saturn 1B rocket launch vehicle to life the *Apollo* spacecraft portion of the Apollo-Soyuz Test Project (ASTP) for rendezvous with the Soviet *Soyuz* spacecraft. This photograph, taken at dawn, shows the *Apollo/Saturn 1B* vehicle in place during a countdown demonstration test. The actual launch took place July 15, 1975 with astronauts Vance Brand, Deke Slayton, and Thomas Stafford.

maneuvering and preparation, the two spaceships docked in orbit (12:09 p.m. on July 17, 1975), 140 miles high.

Aboard the Apollo was Thomas Stafford (commander), Donald Slayton, and Vance Brand. Stafford had been in Space three times, twice on Gemini flights and on *Apollo 10*, and was considered an expert in rendezvous and docking maneuvers that would be required in the joint U.S./U.S.S.R. effort. Slayton, one of the original seven astronauts, at the age of fifty-one, became (at that time) the oldest man to venture into space. Brand was a space rookie.

The two-man crew of the *Soyuz* spacecraft was Alexei Leonov (commander), who had pioneered spacewalking in 1965, and cosmonaut Valery Kubasov, who had previously flown on the 1969 *Soyuz 6* spacecraft mission.

The two spacecraft were linked by a chamber that had docking ports at each end. This docking module was carried into orbit mated to the *Apollo CSM* (Command Service Module).

Three hours after docking, astronaut Stafford opened the hatch of the docking module connecting with the *Soyuz* spacecraft.

"Come in here," Stafford said in Russian to cosmonaut Leonov, who floated through the hatchway.

"Glad to see you," Leonov replied in English.

The two shook hands, cementing their personal friendship and symbolizing a new era of cooperation between the world's two leading space powers.

During the two days the spacecraft were docked, the crews conducted numerous joint scientific experiments, overcoming language, operating, and hardware differences, thereby proving the feasibility of international cooperation in Space.

The three astronauts and two cosmonauts also shared meals, exchanged gifts, and visited each other's spacecraft. The crews had already learned each other's language. When communicating to each other during the joint mission, the astronauts would speak Russian and the cosmonauts would speak English. The American astronauts were introduced to borscht (beet soup), jellied turkey, and black bread.

RETURNING TO EARTH

On July 19, after forty-four hours of being docked with each other, the *Apollo* and *Soyuz* spacecrafts separated and began to make their separate ways back to Earth. On July 21, the Soviet crew aboard the *Soyuz* returned to Earth. The American crew stayed in orbit for three more days, splashing down on July 24.

Comrades in Space. The space crews from the Apollo-Soyuz Test Project were astronauts Thomas Stafford (standing), Donald Slayton and Vance Brand, and cosmonauts Alexei Leonov (standing) and Valeriy Kubasov.

Astronauts and Cosmonaut Together in Space. During July 17 and 18, 1975, the three astronauts and two cosmonauts aboard the linked *Apollo* and *Soyuz* spacecraft visited one another four times. The crews exchanged visits and gifts and shared a meal aboard the *Soyuz* spacecraft. Shown here are astronaut Donald Slayton, cosmonaut Aleksey Leonov and astronaut Thomas Stafford.

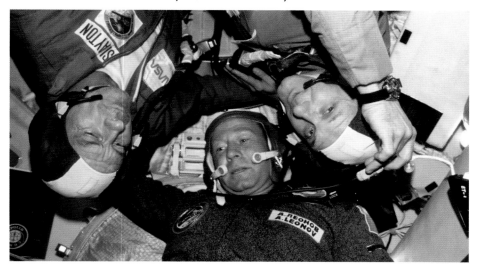

Artist Impression of Apollo-Soyuz Linkup. This artist for the Apollo-Soyuz Test Project shows the docking approach of an Apollo spacecraft to a Soviet *Soyuz* spacecraft. The *Apollo Command Service* module appears on the left; the Docking Module is in the center, and the *Soyuz* spacecraft on the right. Soviet and American engineering design teams devised a system to interface the Docking Module with the existing Soviet docking system. The spacecraft remained linked in Earth orbit for forty-four hours.

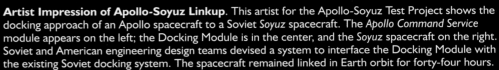

A Meal Aboard the Soyuz Spacecraft. Astronauts Thomas Stafford and Donald Slayton take advantage of the cosmonaut's hospitality and sample their host's space food in the *Soyuz* spacecraft. The containers hold borscht (beet soup) over which vodka labels had been pasted. This was the crews' way of toasting each other.

American Astronaut Vance Brand. Astronaut Vance D. Brand, command module pilot of the *Apollo* spacecraft crew at the controls of the Apollo Command Module.

Apollo-Soyuz Emblem and Stamp. Shown is the official emblem of the joint U.S./U.S.S.R. space mission and a ten-cent stamp, issued in 1975, commemorating the spacecraft's link-up.

The Apollo-Soyuz Test Project

Soviet Cosmonaut Valeriy Kubasov.
Cosmonaut Valeriy N. Kubasov, engineer on the
Soviet crew, is photographed in the *Soyuz* spacecraft.

Apollo-Soyuz Mission Control Room. An overall view
of the Mission Operations Control Room during the U.S./
U.S.S.R. Apollo-Soyuz Test Project docking in Earth orbit
mission. The flight director is seated at his console in the
right foreground, watching the large television monitor which
shows a view of the *Soyuz* spacecraft as seen from the *Apollo*
spacecraft during rendezvous and docking maneuvers.

SPACE SHUTTLE

NASA'S SPACE SHUTTLE IS A REUSABLE VEHICLE LIFTED INTO ORBIT BY ITS own rocket engines and two large rocket boosters. Once empty the boosters fall back to Earth, where they are recovered from the ocean and reused. The Space Shuttle Orbiter is a vehicle as big as a medium-sized airliner. It can stay in orbit for several weeks and carry up to seven astronauts.

Five Space Shuttles were built. They were the most complicated machines that humans have ever built. It is made up of more than 2.5 million parts, contains 230 miles of electrical wiring, and has 27,000 heat-resistant tiles covering its underside.

The Space Shuttle, or as it is more formally referred to, the Space Transportation System (STS), has been described as a Space-going "truck" for hauling cargoes into orbit and returning them safely to Earth. In a sense it is a "truck," but it is also the most sophisticated and versatile spacecraft every developed.

Operating as much more than a transportation system, the Space Shuttle has the combined ability of an aircraft, a launch vehicle, and a spacecraft. The Space Shuttle is comprised of four major components: the airplane-like Orbiter, three main engines, a large external fuel tank, and two reusable solid rocket engines. Of these components, only the external tank is expendable.

The Space Shuttle can be described in the following manner: it is launched as a rocket, orbits the Earth as a spacecraft, and returns to Earth as a glider to a runway landing.

COLUMBIA'S FIRST FLIGHT

It had been six long years since the United States had sent a manned launch into Space. However, this all changed April 12, 1981, as an intense roar filled the air. Space Shuttle *Columbia* (STS-1), the first shuttle to be successfully launched, lifted off from Launch Pad 39 at Kennedy Space Center with astronauts John Young and Robert Crippen aboard. Cheers of jubilation was shared by the thousands of spectators watching the launch, celebrating the renewal of "man in space" and the next generation of Space exploration. On this Space mission, the duration of the flight was two days, six hours, and twenty minutes. The *Columbia* orbited the Earth thirty-six times.

First Space Shuttle Launch. Space Shuttle *Columbia* blasts into Space on April 12, 1981. Never before had a new spacecraft carried humans on its first flight into Space.

Once in orbit, commander Young and pilot Crippen benefited from *Columbia's* large cabin, which was much roomier than previous Space capsules. There was even a second deck beneath the cockpit, with food storage, a private washroom cubicle, spacesuit racks, and an airlock. Two days later, *Columbia* plunged back into the Earth's atmosphere and glided to a successful landing.

For the first time, NASA had a vehicle that could deliver humans and cargo into Space, and then be refurbished for further missions.

THE ORBITER

The most important section of the Space Shuttle is, of course, the Orbiter, designed not only to carry a maximum 65,000 pounds of cargo into low Earth orbit, but also to support up to seven people in Space for several weeks. It is about the same size as a DC-9 commercial airliner, some 120 feet long, with a wingspan of eighty feet and weighing about 150,000 pounds. Most of the structure is aluminum below an outer layer of thermal protection. Moving from the nose to the rear, there is a flight cabin and crew quarters, followed by the 60- by 15-foot box-like cargo, or payload, bay. At the rear there are three main engines plus auxiliary engines for in-orbit maneuvering.

ASSEMBLING THE SHUTTLE

The assembly of the Space Shuttle is done in stages, and preparation for a flight begins many months in advance. At the Orbiter Processing Facility, the Orbiter is serviced, prepared for the next mission, and towed to the Vehicle Assembly Building (VAB). A massive building that was once for the Apollo/Saturn V missions to the Moon, the VAB is now used to assemble Space Shuttle components. The solid rocket boosters, the huge external fuel tank, and the Orbiter are assembled on top of the Mobile Launch Platform (MLP). After the assembly is complete, the MLP is picked up by the Crawler Transporter and begins its long voyage to the launch pad. At the launch pad, routine testing is conducted until the Space Shuttle is ready for liftoff.

Slow Trip to the Launch Pad. A full Space Shuttle stack rolls out of the Vehicle Assembly Building (VAB) for the slow ride to the launch pad. It is mounted on the Mobile Launch Platform and carried by the huge Crawler Transporter. The Orbiter *Enterprise* is mated with the external fuel tank and solid rocket boosters.

Orbiter Discovery in the VAB. In the Vehicle Assembly Building (VAB), the Orbiter *Discovery* (viewed from behind the Space Shuttle's Main Engines) will be raised to a vertical position in order to be mated with the external fuel tank.

SPACE SHUTTLE LIFTOFF

The Space Shuttle uses five engines to lift off the Earth's surface. The main thrust is created by three engines in a triangular formation mounted on the tail end of the Orbiter itself. During liftoff, the main engines exert 1.1 million pounds of thrust.

The external tank is filled with liquid hydrogen and oxygen, and is used to fuel the main engines. The fuel tank and the Orbiter separate when the tank runs out of fuel at about sixty-eight miles from the Earth's surface.

Strapped to the side of the external tank are two solid rocket boosters. The boosters are fueled by a solid-fuel mixture, which provides the Space Shuttle with seventy-five percent of its liftoff power. The two rockets are detached as they run out of fuel. Approximately two minutes into the mission, and about twenty-eight miles in altitude, they parachute back to Earth and are recovered by special recovery ships. They are then towed back to Kennedy Space Center where they are restored for future use.

THE SPACE SHUTTLE IN ORBIT

Once the Space Shuttle Orbiter is in orbit, the longest it can maintain orbit is eighteen days. The time is predetermined according to the missions assigned to the flight. During the astronauts' time in Space, missions may require they deploy satellites from the Shuttle's cargo bay, repair costly satellites in Space, retrieve satellites from Space, or deliver personnel and supplies to an orbiting Space Station.

In 1982 a Space Shuttle mission deployed two commercial satellites. This was indicative of the Space Shuttle's future as a dependable means to successfully deploy commercial satellites and become a profitable venture.

On February 20, 1986, the Soviet Union launched into orbit the *Mir Space Station*, which had a large main compartment, two crew cabins, and multi-docking modules. The Space Shuttle *Atlantis* docked with Mir for the first time in June 1995. For the next three years, several Space Shuttle Orbiters docked at this Space Station.

Placing a Satellite in Orbit. The *Long Duration Exposure Facility* was released with experiments requiring lengthy exposures to Space. Visible behind the arm's wrist joints is Florida.

Birds Scatter as Columbia Leaves the Launch Pad. Startled birds rise in flight from the waters surrounding Launch Pad 39B as the Space Shuttle *Columbia* roars skyward on mission STS-75.

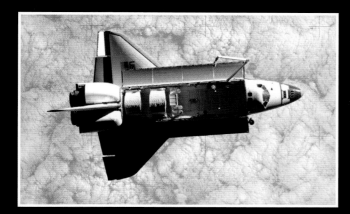

Space Shuttle Challenger in Orbit. This beautiful picture shows the *Challenger* in orbit against a backdrop of clouds (1983).

The Human Satellite. Astronaut Bruce McCandless II looks like Buck Rogers right out of the science fiction pages as he participates in a historic ExtraVehicular Activity (EVA), a short distance away from the Earth-orbiting Space Shuttle *Challenger* (STS-11). This Space Walk represented the first use of a nitrogen-propelled, hand-controlled device, called the Manned Maneuvering Unit (MMU), a backpack propulsion device, which provided a new degree of freedom for astronauts in orbit.

Capturing a Faulty Satellite. Docking with a Space Station to bring supplies, equipment, or a new crew, servicing a Space telescope or capturing a faulty satellite are routine activities involved in the Space Shuttle program. Astronaut Pierre Thuot is seen on the end effectors of the Robot Arm, attempting to capture the Intelsat VI satellite, in 1992.

On April 24, 1990, the *Hubble Space Telescope* was loaded into the cargo bay aboard the Space Shuttle *Discovery* and successfully launched into Space. The *Hubble Telescope* now orbits 381 miles above the Earth.

In December 1993, a Space Shuttle mission was launched to repair the *Hubble Space Telescope* in Space. Additional repair missions were sent in the following years.

Since the late 1990s, the majority of the Space Shuttle flights have been involved with taking crew and equipment to and from the *International Space Station*.

SPACE SHUTTLE REENTRY

Reentry into Earth's atmosphere from Space must be a hair-raising experience. As the Space Shuttle Orbiter heads home, it must transform itself from spacecraft to glider, and lose most of its 17,500-mph orbital velocity. The Orbiter must turn in Space, fire-braking rockets, and then plunge into the atmosphere for a fiery 3,000°F ride until friction slows it to orthodox flying speeds. Then, with engines silenced, it glides toward the welcome sight of the landing strip. A chase plane escorts the Orbiter to its landing site.

TOUCHDOWN!

One of the chief hazards faced by the Orbiter during landing is poor weather, and hurricane-prone Florida sometimes has adverse conditions over the Cape Canaveral area. When this occurs, the Orbiter is directed to land at Edwards Air Force Base in California, where the weather is invariably good.

During the final moments of landing, the Orbiter is joined by a T-38 chase airplane. The T-38 pilot repeatedly calls speed and altitude readings to the Orbiter to help them land the world's heaviest glider. A few seconds after touchdown, the main drag chute is deployed to help slow the Orbiter from its landing speed of around 250-mph. The Orbiter is capable of completing its landing sequence on autopilot, but Orbiter crews generally prefer to switch to manual override for the last crucial seconds.

Night Landing. At the end of the missions, the Space Shuttle Orbiter reenters the atmosphere traveling at about Mach 25 (twenty-five times the speed of sound). The atmosphere acts like a brake, slowing down the Orbiter to a safe landing speed. Shown is a night landing of the Space Shuttle *Columbia* at the Kennedy Space Center in 1996.

Crew of the Challenger. The crew of the *Challenger* for Space Shuttle mission 51-L poses for the traditional crew portrait some months before liftoff. From left to right, back row, are Ellison Onizuka, Christa McAuliff, Greg Jarvis, and Judy Resnik. Front row: Mike Smith, Dick Scobee, and Ron McNair.

CHALLENGER AND COLUMBIA DISASTERS

On January 28, 1986, an unbelieving world witnessed the Space Shuttle *Challenger* explode during a routine launch. The cause of the accident was the failure of the pressure seal in the aft field joint of the right-hand solid rocket booster, due to a faulty design.

Aboard the doomed *Challenger* were Francis Richard Scobee (Mission Commander), Michael John Smith (Pilot), Ronald Erwin McNair (Mission Specialist), Ellison Shoji Onizuka (Mission Specialist), Judith Arlene Resnik (Mission Specialist), Gregory Bruce Jarvis (Payload Specialist), and Sharon Christa McAuliffe (Payload Specialist). Teacher Christa McAuliffe was the first "ordinary American citizen" to fly on a Space Shuttle mission.

On the day the *Challenger* seven died, President Ronald Reagan said, "They slipped the surly bonds of Earth to touch the face of God."

The launch of the Space Shuttle *Columbia*, on its twenty-eighth mission, on January 16, 2003 appeared to be entirely routine and successful as the crew performed their mission. Less than two minutes after launch, a piece of insulating foam from the external fuel tank fell off and struck the Orbiter's left wing and set in train a sequence of events leading to the breakup of *Columbia* on reentry fifteen days later.

The *Columbia* was traveling at approximately 12,500 mph when it broke up over Texas at an altitude of 207,000 feet, spreading debris over a wide area of Texas and into Louisiana. Aboard the lost *Columbia* were Rick Husband (Mission Commander), William McCool (Pilot), David Brown (Mission Specialist), Laurel Clark (Mission Specialist), Michael Anderson (Payload Commander), and Illan Ramon (Payload Specialist). In 2003, seven asteroids orbiting the Sun between Mars and Jupiter were named after the *Columbia* crew.

INTERNATIONAL SPACE STATION

AMERICA'S INVOLVEMENT WITH SPACE STATIONS HAVE occurred during three space programs: *Skylab*; Shuttle-Mir; and the *International Space Station*.

Skylab, America's first experimental space station (1973-1974), was discussed in "Chapter Eight."

The Shuttle-Mir Program involved the Space Shuttle transporting astronauts and supplies to the Russian *Mir Space Station* (1995-1999).

The *International Space Station* (1998-present) is an achievement of sixteen nations working together to construct a permanent station on the fringes of the last frontier — Space.

SHUTTLE-MIR PROGRAM

Mir was the first permanent space station, a unique complex designed for expansion. By the time the station was complete it looked like a product of a bizarre engineering experiment in zero gravity. The station's modular design resulted from the payload limitations of the Proton rocket. It eventually comprised of six modules, including a docking module for use by the Space Shuttle. All of the Soviet modules were delivered by Proton rockets and docked automatically.

Composition of the Station

A Proton rocket at Baikonur, Kazakhstan, launched the base module of the *Mir Space Station* on February 19, 1986.

Added to the base module were the Kvant 1 Astrophysic Module on November 26, 1989; the Kristall Technological Module on May 31, 1990; the Spektr Science Module on May 20, 1995; and the Priroda Science Module on April 23, 1996.

A *Soyuz* spacecraft is permanently attached to *Mir* as an escape vessel for reentry to Earth.

The three-section base module contained the living quarters. At the front was a separate airlock/five-port docking node. At the rear was the service section with thrusters, main engine, tanks for consumables, and communications.

Departing from Mir. The Space Shuttle *Atlantis* departs from the *Mir Space Station* (1995).

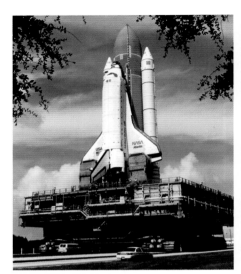

Shuttle to the Mir Space Station. Space Shuttle *Atlantis* rolls out to Launch Pad 39A on August 20, 1996 in preparation for launch of STS-79 on the fourth *Mir Space Station* docking mission. *Atlantis* will return astronaut Shannon Lucid to Earth after her record-breaking stay by an American on the Russian space station. Lucid had just completed twenty-one weeks in space.

Docking at Mir. Space Shuttle *Discovery* approaches the *Mir Space Station* (1998).

A View from Mir. Fish-eye view of Space Shuttle *Atlantis* as seen from the *Mir Space Station* (1995).

Space Shuttle-Mir Linkup

The Shuttle-Mir Program was a joint effort between the United States and Russia designed to conduct scientific experiments from Space and construct space stations. The program's focal point was the Russian *Mir Space Station*.

The American Space Shuttle extended the capability of Mir by providing large-haul capacity and the Mir Space Station provided living and working quarters. A series of Shuttle missions transported NASA astronauts to Mir where they served as members of the crew. Before the astronauts were sent to Mir, they learned to speak the Russian language.

The first Space Shuttle *Atlantis* (STL-71) was launched from Launch Complex 39A at the Kennedy Space Center on June 27, 1995, with five astronauts and two cosmonauts, a Mir relief crew, aboard. The flight was to be the first in a planned series of Shuttle-Mir dockings over a three-year period. The *Atlantis* carried a docking mechanism that was especially constructed for the task. Two days after launch, the *Atlantis* docked with the Mir's Kristall Module docking port. For the first time in twenty years, an American and a Russian spacecraft were linked in Earth orbit.

After the docking hatches were opened, the crews of the *Atlantis* and Mir greeted each other, exchanged toasts and gifts, and set about conducting joint activities and experiments for the next five days. This was the first time that American astronauts had the chance to study the effects of extended weightlessness on the human body.

Over the years nine Space Shuttles visited the *Mir Space Station*. The Shuttles delivered fresh supplies, hardware components, a relief crew, and conducted scientific experiments.

Problems Aboard the Mir

In the late 1990s Mir began to have problems. In 1997 the station suffered a serious fire and a collision with a Progress spacecraft that caused partial depressurization. Scientific work ceased and the last team abandoned ship on August 27, 1999. It had become obvious that the station's time was over.

NASA astronaut Jim Lovell, hero commander of the *Apollo 13*, said, "Mir has done an exceptionally fine job. Now it's time to give it a very respectful retirement."

On March 23, 2001, Mir was de-orbited where it fell into the Earth's atmosphere and much of it burned up. Several tons of the station fell harmlessly into the Pacific Ocean. The *Mir Space Station* had stayed in orbit for 5,511 days.

Over its lifetime, thirty-one manned spacecraft docked with the *Mir Space Station*, including cargo spaceships, over one hundred different crews from twelve different nations (125 visitors in total), seventeen visiting expeditions, and twenty-eight long-term crews. Mir completed just over 89,000 orbits at an altitude of 240 miles above the Earth's surface. Keeping a human environment orbiting in the vacuum of space for thirteen years was one of the great technological achievements of the twentieth century.

Some Benefits from Mir

Mir showed that spacemen and spacewomen from different countries could live and work together in Space for long periods of time. Cosmonauts were required to make many space repairs and became expert mechanics, solving problems that no one had ever thought of. Cosmonaut Valeri Polyakov spent fourteen months in space, a record that will probably stand for some time.

Many scientific experiments carried out on Mir provided vital data. Alloys and vaccines made impossible by gravity were manufactured in space. Even food was grown and harvested in space. Fish survived in an aquarium.

Mir showed that people can live in space for a year — long enough to plan for follow-up scientific work on the International Space Station, and long enough for a round-trip to Mars. The lessons learned from Mir were later put to use in the development of the International Space Station.

INTERNATIONAL SPACE STATION

The *International Space Station* (ISS) is the largest structure ever to fly in space. The program involves sixteen nations: the United States, Russia, Canada, Japan, Brazil, Belgium, Denmark, France, Germany, Italy, the Netherlands, Norway, Spain, Sweden, Switzerland, and the United Kingdom. The ISS is controlled by the United States and Russia equally.

The *International Space Station* is an achievement of nations working jointly to construct a permanent station in space. From Earth orbit, the orbiting laboratory allows researchers to conduct experiments in a near gravity-free environment, astronomers to observe an unobstructed view of the universe, and humans to live and work in space for extended periods. It is an excellent platform for observing the Earth. From their high vantage point, scientists study weather patterns, land usage, the destruction of rain forests, and the spread of deserts.

The ISS is made up of several modules connected together. It is something like a giant model kit being assembled piece-by-piece 224 miles above the Earth's surface. ISS modules can range from complete research

International Space Station

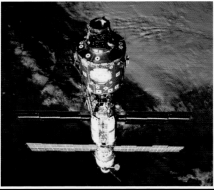

First Two Components of the ISS. In December 1998, the Space Shuttle *Endeavor* astronauts released Unity (right) from the cargo bay and connected it to the *International Space Station* component Zarya, launched earlier on a Russian Proton rocket. The six-sided central connector, Unity, will link future components to the ISS.

Third Component Added to the ISS. In September 2000 the astronauts of Space Shuttle STS-106 linked the second Russian component, the service model Zvezda, to the station.

Solar Panels Added to the ISS. A 2001 view of the *International Space Station*, photographed from the aft deck of the Space Shuttle *Discovery*. The ISS now contains several modules and solar panels for power.

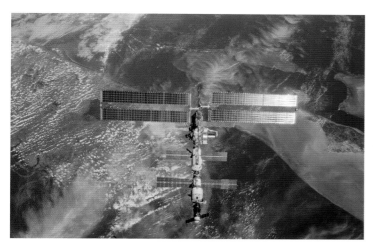

A Space View of Miami. The *International Space Station* has been growing steadily since 1998, as new modules, or parts, have been added. This is how it looked in December 2001. The ISS is back-dropped over Miami, Florida, in this view photographed by Commander Dominic Gorie aboard the Space Shuttle *Endeavour*.

224 Miles Above the Earth. The *International Space Station* continues the largest scientific cooperative program in history, drawing on the resources and scientific expertise of sixteen nations.

The International Space Station. A view of the *International Space Station* as it moves away from Space Shuttle *Endeavour*. Earlier the shuttle crew conducted several days of cooperative work onboard the orbiting station.

Crew Inspects Endeavour's Payload Bay. Space Shuttle crew at the Kennedy Space Center inspecting a module that they will carry to the *International Space Station* on a future flight.

Space Shuttle Docked at the ISS. This view shows how the *International Space Station* will look once it is completed.

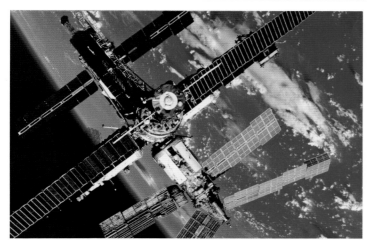

laboratories to small parts. The modules or parts, which must fit in the cargo bay of a Space Shuttle, are hauled into space by American Space Shuttles and Russian boosters. These modules are put together in space by spacewalking astronauts and cosmonauts aided by robot arms.

The sprawling Space Station complex is made up of numerous cylindrical sections, or modules, containing laboratories, workshops, and living quarters for a crew of six. Russian-built modules include the U.S.-financed Zarya base module, the Zvezda service module, and the Pirs docking module. U.S. modules included the Unity connector module, the Destiny research lab, and the Quest airlock. Canada provided the station's robotic arm. The European Space Agency provided the Columbus laboratory and Japan the Japanese Experiment Module. The *International Space Station* is 356 feet long and 290 feet wide, weighs about 450 tons, and is about the size of a football field.

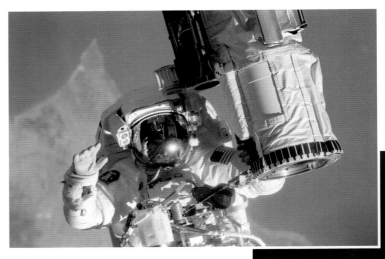

A Construction Astronaut at Work.
Holding onto the end effectors of the robot arm on the Space Shuttle *Atlantis*, astronaut Michael L. Gernhardt performs work on the *International Space Station*. The jutting peninsula in the background is Cape Kormakiti on the north central coast of Cyprus and the water body to the left of the cape is Morphu Bay.

Space Shuttle Discovery. A view of the Space Shuttle *Discovery* as it approaches the *International Space Station* during docking operations. The Harmony node, shown in the cargo bay, is being delivered to the orbiting station.

Russian Rocket Launch to the ISS. A Russian Soyuz rocket launches from the Baikonur Cosmodrome in Kazakhstan carrying an American and Russian crew to the *International Space Station*. Russian rockets and American Space Shuttle's transport crew and supplies to the station.

Continuing Space Station Construction. A fish-eye view of the Space Shuttle *Endeavour* (STS-118) as it lifts off from Launch Pad 39A on the twenty-second shuttle flight to the *International Space Station*. The Shuttle delivered a module and other parts to the orbiting station.

Crews are taken to and from the ISS by Space Shuttles and the Russian *Soyuz* spacecraft. Supplies and propellants are delivered by Space Shuttle flights and the unmanned robotic Russian *Progress M* spacecraft. The first resident crew arrived in November 2000. Since then, the ISS has been permanently occupied. Resident crews usually stay in orbit for six months.

The Moon from Space. Framed by components of the *International Space Station*, a full moon is visible in this view above Earth's horizon and airglow, photographed by a crewmember on the Space Station *Endeavour* while it was docked with the orbital complex.

Back from the International Space Station. Under a blue sky washed with clouds, *Endeavour* has wheels down for a landing on the runway at Kennedy Space Center, completing a thirteen-day mission to the *International Space Station*. The Space Shuttle crew had just installed a new gyroscope, an external spare parts platform and another truss segment to the expanding station.

Astronauts Working in Space. Astronaut Daniel Tani (top center), during a six-hour, 33-minute spacewalk with astronaut Scott Parazynski (out of frame), performs construction work on the *International Space Station*. The Moon is visible at lower center.

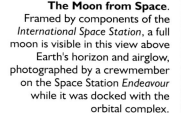

Walt Disney World's Magic Kingdom Welcomes Visitors from Space. A ticker-tape parade officially welcomes toy space ranger Buzz Lightyear home from space. The twelve-inch tall action figure spent more than fifteen months aboard the *International Space Station*, returning to Earth aboard Space Shuttle *Discovery* with the shuttle crew. Lightyear's space adventure, a collaboration between NASA and Disney Parks, was intended to share the excitement of space exploration with students around the world and encourage them to pursue studies in science, technology, engineering, and mathematics. NASA astronauts were featured in the procession.

THE HUBBLE SPACE TELESCOPE

THE ULTIMATE TELESCOPE FOR ASTRONOMERS SEEKING PIN-SHARP VIEWS OF the depths of the universe is the *Hubble Space Telescope* (HST). On April 15, 1990, the HST, named for astronomer Edwin P. Hubble, was loaded into the cargo bay of the Space Shuttle *Discovery* (STS-31) and launched from Pad 39. The shuttle crew deployed the telescope at 350 miles above the Earth where it would orbit and gather data. Its mission was to take a closer look at our Solar System, the Milky Way and other galaxies, and to gaze back in time into the farthest reaches of the universe. The *Hubble Telescope* sees further and clearer than any other telescope on Earth since there is no atmospheric interference. It is the largest, most complex, and most powerful observatory ever deployed in Space.

The *Hubble Space Telescope* is a joint NASA-ESA (European Space Agency) project. It was designed for a fifteen-year lifetime, with periodic maintenance visits from Space Shuttle astronauts. The 94.5-inch primary concave mirror collects light from distant objects, which is sent to various instruments. Fine guidance sensors lock onto guide stars to ensure extremely high pointing accuracy.

Soon after deployment, it was discovered that the main mirror of the Hubble was faulty, so on December 2, 1993, the Space Shuttle *Endeavour* (STS-61) was launched on an eleven-day mission to repair the telescope in Space. The *Endeavour's* crew successfully repaired the telescope, making it once again capable of gathering data from up to fourteen billion light years away.

Since then, there have been several other servicing missions:

✦ The February 19, 1997, service mission added two new instruments and made repairs.

✦ The December 1999 service mission installed a new computer and replaced Rate Sensing Units.

✦ The March 2002 mission installed an advanced camera and replaced instruments, gyros, and solar panels.

✦ The 2008 mission placed the Hubble on two-gyro operating mode, extending its lifetime.

✦ In May 2009, the Space Shuttle *Atlantis* (STS-125) was sent to upgrade the telescope.

The *Hubble Telescope* is a big satellite, roughly the size of a railroad tank car, over forty-three feet long, with a diameter of fourteen feet, and weighs about twelve and a half tons. Two large solar arrays, made up of thousands of solar cells, provide the telescope with electricity to keep its batteries charged. Attached to its eyepiece are two cameras; one that looks in great detail at a small area of space, and one that focuses on much larger areas or objects. Other instruments analyze the infrared waveband and characteristics of light.

The Hubble orbits the Earth about once every ninety-five minutes. Astronomers from dozens of countries use the Hubble, operating it by remote control.

Launch of the Hubble Space Telescope. Its cargo bay packed with the *Hubble Space Telescope*, Space Shuttle *Discovery* left Launch Pad 39A at Kennedy Space Center on April 24, 1990. Amid colorful flames and a rumble of thunder, the crown of the international astronomical research community was beginning its journey. The following day, the shuttle crew, using the mechanical arm, placed Hubble into orbit. Prior to its release, the Hubble's solar panels were extended. These gather the energy that makes the various systems onboard function.

Routine Servicing Mission. Astronaut Kathryn Thornton, tethered to the Shuttle's robot arm, removes a damaged panel from the *Hubble Space Telescope* in its first servicing mission, December 1993. Hubble was the first major satellite designed specifically for in-orbit servicing.

First Repair in Space. In one of the most daring events in the history of spaceflight, in December 1993 the crew of Space Shuttle *Endeavour* (STS-61) repaired the *Hubble Space Telescope*. Shown here are astronauts Story Musgrave, tethered to the Shuttle's robot arm, and Jeffrey Hoffman, in the Shuttle's cargo bay, in the process of fixing the crippled Hubble.

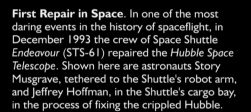

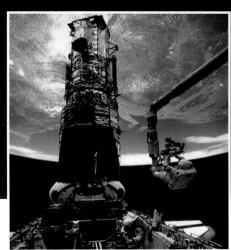

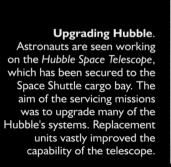

Upgrading Hubble. Astronauts are seen working on the *Hubble Space Telescope*, which has been secured to the Space Shuttle cargo bay. The aim of the servicing missions was to upgrade many of the Hubble's systems. Replacement units vastly improved the capability of the telescope.

The Hubble Space Telescope Hovers Over Earth. The *Hubble Space Telescope* was carried into orbit on the Space Shuttle *Discovery* (STS-31) April 24, 1990. The Hubble has beamed hundreds of thousands of images back to Earth and transformed the way scientists look at the universe.

The Hubble Space Telescope

Routine Service Mission on the Hubble. Astronauts install a Fine Guidance Sensor into a protective enclosure in the Space Shuttle's cargo bay. These two astronauts performed the second of three space walks to service the *Hubble Space Telescope* on the STS-103 mission in December 1999.

Hubble Space Telescope and the Moon. A Full Moon, as photographed by the crew aboard the shuttle *Discovery*, is shown about three hours before a rendezvous with the *Hubble Space Telescope* (HST), which appears as a tiny dot above and to the left of the Moon. The picture was taken with an electronic still camera (December 1999).

Hubble Space Telescope in Orbit. The *Hubble Space Telescope* has the benefit of being above the atmosphere, enabling it to give us a much clearer view of the heavens — the pictures it has sent have been truly stunning. The Hubble has photographed electrifying views of distant stars and galaxies and revealed new clues about the origins of the universe.

HUBBLE OBSERVATIONS

For over two decades the *Hubble Space Telescope* has enabled us to see and understand our universe more clearly. It has also helped us to place our home and our lives in a wider context, balancing our concept of the vastness and permanence of planet Earth with the realization that we actually live on a rather small planet and that half the planets in our Solar System are much bigger. The Hubble has allowed us to probe the childhood of our universe and witness the formation of galaxies, investigate the 200 billion stars in our own galaxy, see the extraordinary spectacle of star birth and death, and explore objects in our Solar System.

The Hubble has taken over 750,000 exposures and studied over 24,000 celestial objects. Millions each day now access the Internet to witness the beauty and splendor reflected in the Hubble's eye, trained on the far corners of the universe.

The *Hubble Telescope* was designed to make a large number of very different observations over the entire range of astronomical objects, from asteroids and comets in the Solar System to super-massive clusters of galaxies at the edge of the Universe. While the telescope cannot get the detailed images of other planets that the Mariner and Voyager have, it has the advantage of being able to make more frequent and sustained

Hubble Service Mission Getting Ready to Launch. The Space Shuttle *Atlantis* (foreground) sits on Launch Pad 39A and *Endeavour* on Launch Pad 39B at the Kennedy Space Center. For the first time since July 2001, two shuttles are on the launch pads at the same time at the center. *Endeavour* was standing by in the unlikely event that a rescue mission was necessary during *Atlantis'* upcoming mission to repair the *Hubble Space Telescope*, targeted to launch in October 2008.

Atlantis—Ready to Go on Another Mission. The Space Shuttle *Atlantis* on Launch Pad 39A is viewed across the lagoon at the Kennedy Space Center. The *Atlantis* was set for a May 2009 launch on the STS-125 mission to upgrade the *Hubble Space Telescope*.

observations, looking for changes in the atmosphere and temperature.

The Hubble is also being used to probe our own Milky Way. Its superior resolution enables it to see the details of star formation in huge clouds of gas and dust. The images sent home by the Hubble Space Telescope mesmerized the world and thrilled astronomers:

✦ **1992**: The Hubble found evidence for a massive black hole in the galaxy.

✦ **January 1994**: The Hubble recorded comet Shoemaker-Levy 9's impact with Jupiter. It also mapped the surface of Titan (Saturn's largest moon) using infrared wavelengths, and photographed the surface of Pluto, and Charon, a satellite of Pluto.

✦ **1995**: The Hubble photographed star birth in the Eagle Nebula. Hubble also produced a view of the most distant galaxies, up to 10 billion light-years away, and photographed the planet Jupiter on February 13 with images comparable in richness of detail to those from the Viking and Voyager spacecraft missions to Mars, Jupiter, and Saturn.

✦ **July 24, 1996**: The Hubble photographed the dome-shaped eruptive plume of Pele on Io, one of Jupiter's largest Moons.

✦ **1997**: The Hubble checked landing sites for *Mars Pathfinder* space probe, monitored dust storms on Mars, took high-resolution photographs of the lava-covered asteroid Vesta, and produced view of faint comets passing in Earth's general vicinity: comets Hale Bopp, Borrelly, Honda-Mrkos-Pajdusokova, Kopff, Encke, Wirtanen, Hyakutake, and Schwassmann-Wackmann I.

✦ **1999**: The Hubble detected the galaxy containing an energetic gamma-ray burster — the most powerful explosion ever observed.

THE HUBBLE'S SUCCESSOR

The *Hubble Space Telescope* is expected to reach the end of its life in 2013. The Hubble's successor will be the *James Webb Space Telescope* (JWST), named for James Webb, a NASA administrator during the Apollo era.

The JWST will work from a gravitationally stable location known as Lagragian Point 2, some 940,000 miles away, beyond Earth's Moon. The idea is to site the JWST at the point where the pull of gravity from the Earth and the Sun cancel each other.

The JWST will be three times bigger, yet four times lighter, than the Hubble. The design features an ultra-lightweight mirror twenty feet across, made of a beryllium composite rather than glass. The JWST is scheduled to be assembled in 2012 and ready for launch by the European rocket Ariane 5 in 2013. Hopefully, it will be able to see galaxies forming at the very edge of the universe and look for life on planets orbiting other stars.

Glowing Eye Nebula in the Constellation Aquila. In 2000 the *Hubble Space Telescope* (HST) captured the image of the Glowing Eye Nebula in the Constellation Aquila. The HST has revealed the staggering beauty of many other planetary nebulas, including the Eskimo Nebula in the Constellation Gemini, the Carina Keyhole Nebula, the Crab Nebula, the Eagle Nebula, and the Giant Galactic Nebula.

Cat's Eye Nebula in Constellation Draco. The Cat's Eye Nebula (NGC 6543) is one of the most beautiful and complex planetary nebula the *Hubble Space Telescope* has studied. The eye is surrounded by concentric rings, which are spherical bubbles of material ejected at 1,500 yearly intervals. It is 3,000 light years away in the constellation Draco.

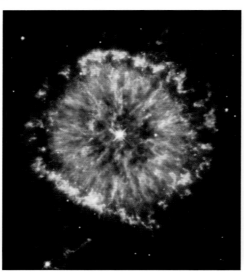

Stellar Stalagmites. The eerie pillars of dark dense gas in the heart of the Eagle Nebula are incubators for new Stars. They protrude from the wall of a dark molecular cloud rather like stalagmites from a cave floor. The Eagle Nebula is 7,000 light-years away and 70 by 55 light-years wide. The image is composed of three shots taken in red, green and blue light with the *Hubble Space Telescope's* wide field, planetary camera.

Spiral Galaxy. This remote galaxy is NGC 4603, just one in a cluster of galaxies in Centaurus. The *Hubble Space Telescope* has been able to pinpoint Cepheid variables in the galaxy, establishing its distance at 108 million light years.

Heart of the Whirlpool. On the *Hubble Space Telescope*'s fifteenth anniversary in April 2005, it returned to the celebrated Whirlpool Galaxy. The two curving arms are shown in striking detail. Clusters of young stars are highlighted in red.

Sirius, the Dog Star. The brightest star in the nighttime sky, close inspection of Sirius shows it is made of two stars. Sirius A in the center of this *Hubble Space Telescope* image is over-exposed so that its tiny companion, Sirius B (dot at lower left), is visible.

Ancient Stars in the Milky Way. A Milky Way globular cluster of ancient stars illustrates the kind of *Hubble Space Telescope* pictures that shed light on the origins of the universe.

THE EARTH AND ITS ARTIFICIAL SATELLITES

THE PLANET EARTH IS THE FIFTH LARGEST IN THE SOLAR SYSTEM AND THE third closest to the Sun. It is a spinning sphere that is bounded by the black vacuum of Space. It is a bright blue and white sphere that spins on its axis once every twenty-four hours. Earth has one natural satellite, the Moon, and over 6,000 artificial satellites.

EARTH: THE BLUE PLANET

Viewed from Outer Space, the Earth is one of the most beautiful of all the known planets. The oceans, which cover seventy percent of its surface, and the clouds, which constantly swirl over the land and sea, reflect the Sun's light so that the planet shines bright and blue against the blackness of Space. The Earth was born 4.6 billion years ago, but almost 1.6 billion years were to pass before signs of life appeared on its surface.

Earth from Apollo 11 Spacecraft. This view from the *Apollo 11* spacecraft shows the Earth, 240,000-miles away, rising above the Moon's horizon. The lunar terrain pictured is in the area of Smyth's Sea on the nearside. The two worlds are very different — the Earth is bright and bustling while the Moon is dull and lifeless.

Satellites are small heavenly bodies that orbit larger bodies, such as the planets. The Earth has one natural satellite, the Moon, and because it is only 240,000 miles away, and clearly visible from Earth with the naked eye, it is the most familiar of all the heavenly bodies. The Moon creates no light of its own. It shines by reflecting the light of the Sun. The Moon orbits the Earth in just over twenty-nine days, so this is how long it takes for each cycle of the Moon's phases.

ARTIFICIAL SATELLITES

On October 5, 1957, to the surprise of the United States and the World, the Soviet Union launched *Sputnik 1*, the first artificial or man-made satellite into orbit. *Sputnik 1* was nothing more than a beeping radio transmitter, but many practical applications have since been found for artificial satellites. Today, there are many satellites in Earth orbit. Satellites relay television programs, telephone calls and data, observe changes in the environment, and provide accurate location data to GPS receivers, which have a wide range of military and civil applications. They

Earth. This view of Earth was taken from the *Apollo 17* spacecraft. Earth is called the "blue planet" because three-quarters of its surface is covered with water, made up of six major oceans and many seas, lakes and rivers. The Pacific Ocean, the biggest ocean in the world, is much larger than all the land on the Earth put together. This photograph reminds us that the Earth's entire biosphere exists in the thin veneer where land, ocean, and atmosphere meet.

are our eyes and ears in space and monitor how our planet is doing. They report on the state of the atmosphere and provide detailed maps of the landmasses and oceans. These space-based laboratories have even helped scientists understand our planet's geological formation and history. Satellites have military uses such as launching ballistic missiles and listening to enemy communications. The modern world would barely function without satellites.

An artificial satellite's journey into Space usually involves several phases. First, the launching rocket places its payload into low Earth orbit. Then a rocket burns to send the satellite out along an elliptical transfer orbit. Finally another rocket motor, built into the satellite or connected to it, puts the craft permanently into the planned orbit. Other artificial satellites are carried into Space in the cargo bay of the Space Shuttle, where they are released with motors to kick them into their final orbit.

Since *Sputnik 1* shook the world, more than 6,500 artificial satellites have been launched. Although most of these have been inserted into various orbits around the Earth, several hundred robotic ambassadors have been dispatched throughout the Solar System. Many of the Earth orbiting satellites fall into five categories: Communications, Navigation, Earth Resources, Meteorology, and Military.

Communications Satellites

Telephone calls, television broadcasts, and the Internet can all be relayed by communications satellite. These satellites connect distant places and make it possible to communicate with remote areas. Many are in geostationary orbit, but so great is the demand for communications that this orbit has become crowded. Since the 1990s, fleets of satellites have been launched into low-Earth orbit to carry signals for the growing number of cellular phones.

The history of communications satellites goes all the way back to December 1958, when the United States launched an Atlas rocket into orbit to beam a taped message from President Eisenhower to radio listeners.

The *Signal Communication by Orbiting Relay Equipment* (SCORE) satellite, launched in 1960, relayed radio waves bounced up from Earth. The 52-inch sphere could pass voice, teletype, and fax data from its origin on Earth back down to another location.

Also in 1960, NASA launched *Echo 1*, a one hundred-foot aluminum-coated balloon that acted as a relay satellite. It allowed the first two-way communication via satellite. Bell Laboratories used *Echo 1* to send the first transoceanic satellite message to Paris, France a few days after the satellite began orbiting Earth.

The Delta (Thor-Delta) rocket at Cape Canaveral carried into space the world's first commercial satellite, AT&T's pioneering TV communications relay unit *Telstar 1* on July 10, 1962, which introduced the TV phrase: "Live via satellite."

Telstar 1 transmitted the first live television signal across the Atlantic Ocean on its first day of operation. Scientists at the receiving station in France beamed with delight when they saw a picture of the American flag on a television screen and heard "The Star-Spangled Banner." The transmission left the United States, traveled to France, and then back again. The enterprise fueled imaginations worldwide and was soon followed by many more communications satellites with names such as *Syncom*, Intelsat's *Early Bird*, and *Comsat*. The *Telstar 1* remains in orbit around the Earth; however, it was shut down February 21, 1963.

The world's first geostationary satellite for commercial traffic was the Intelsat's *Early Bird*, launched in April 1965.

International Telecommunications Satellite (Intelsat) owns and operates a global system of communications satellites for commercial use. Intelsat satellites, in geostationary orbit, have revolutionized world

Application Technology Satellite (ATS) Launch. Lift off of Atlas-Agena launch vehicle with *ATS-3* spacecraft from Launch Complex 12 on November 5, 1967.

Satellite Retrieval. On November 8, 1984, the Space Shuttle *Discovery* (Mission 51-A) was sent to recover two communication satellites, the *Westar VI* and *Palapa B-2*. Astronaut Dale Gardner is seen flying an MMU during the *Westar's* successful recovery.

Syncom IV Communications Satellite. The first Syncom satellite, *Syncom II*, was launched February 14, 1963, by a Thor-Delta rocket. On July 26, 1963, *Syncom II* was placed in orbit. In 1964, *Syncom III* was placed in a geostationary orbit over the Pacific Ocean. *Syncom III* was used to relay live TV coverage of the 1964 Olympic Games in Tokyo. *Syncom IV* (shown) was placed in orbit by Space Shuttle STS-41 in September 1984.

communications, from placing a telephone call to receiving TV pictures of world events as they happen.

In 1967, the Intelsat II series, consisting of three satellites, was stationed over the Atlantic and Pacific to increase the range of service. A year later, the Intelsat III series, which provided global commercial communications, was capable of handling 1,200 telephone circuits or two TV channels at the same time. Intelsat IV, entered service in 1971, with an expanded capability, and by 1975 the satellites were capable of handling 6,000 telephone calls plus two television channels. By 1985, the Intelsat satellites had a capability of handling 15,000 telephone calls. Newer Intelsat satellites are able to handle several tens of thousands of telephone calls.

In 1975, NASA's *ATS-6* satellite was stationed over India to broadcast education programs to small ground stations.

Satellite television began in 1978, when the Public Broadcasting Station (PBS) first used satellites to distribute programs to its local stations. By the early 1980s, direct-to-home (DTH) satellite receivers were commonplace, especially in rural areas where regular broadcast quality was poor and cable nonexistent. Satellite TV rapidly went global. The *Galaxy* satellite, launched February 9, 1994, weighs 3,800 pounds and can transmit more than two hundred separate TV channels.

Today, businesses, industry, and individuals alike rely on information supplied by satellites. Examples include mobile voice communication via satellite, knowledge based on the Internet, satellite-relayed television programs, remote sensing images from Space, and location data from the Global Positioning System (GPS) satellites.

Communications satellites have become a way of life. The modern world would barely function without satellites, and new launches will need to be made far into the foreseeable future.

Navigation Satellites

To steer an accurate course between two places, a navigator needs to know his or her exact position. For thousands of years, sailors calculated their positions using the Moon, Sun, and the stars. When clouds obscure the sky, however, it is easy to stray off course. Satellite navigation systems have solved this problem. Satellites transmit radio waves that can be detected on Earth even when it is cloudy. As a result navigation is now possible in any weather.

Improved positional accuracy from navigation satellites help ships chart more accurate courses and adds to safety in crowded sea lanes, as well as reducing confusion over the limits of territorial waters. Offshore oilrigs use navigation satellites to ensure they are drilling in the correct position.

Transit was the first satellite navigation system. It was launched in 1964 to improve position location of Polaris nuclear submarines. The U.S. Navy made *Transit* available to civilian users in 1967. In 1978, the U.S. Air Force launched the first GPS (Global Positioning System) satellite. More GPS satellites were launched in the 1980s and 1990s, increasing the number of places at which signals could be received at every minute of the day. Everyone can now benefit from GPS. Motorists looking for the best travel route can now get a GPS readout or printout, scientists can track members of a lion pack, and police can locate a victim who dialed 911. No one with a suitable receiver need now be lost anywhere on Earth. Today, the GPS system has become the most reliable system ever.

Earth Resources Satellites

Satellites that help scientists study Earth's surface are called Earth resources satellites. They can show whether crops are failing or ice caps are melting, and can pinpoint resources such as metal ore or coal. This is possible because the satellites' instruments analyze light and other radiation reflected and emitted from surface features. Each feature, a forest or building for instance, has a different signature of reflected and emitted radiation. Satellites pass regularly over the whole globe, allowing scientists to produce maps that trace how a particular area changes over time.

Remote sensing from space detects features invisible to ground-level observers, and has transformed our knowledge of the Earth. Mining companies use remote sensing imagery to guide excavations. Farmers can accurately estimate seasonal crop yields. Before-and-after pictures

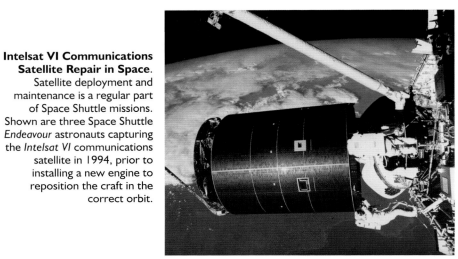

Intelsat VI Communications Satellite Repair in Space. Satellite deployment and maintenance is a regular part of Space Shuttle missions. Shown are three Space Shuttle *Endeavour* astronauts capturing the *Intelsat VI* communications satellite in 1994, prior to installing a new engine to reposition the craft in the correct orbit.

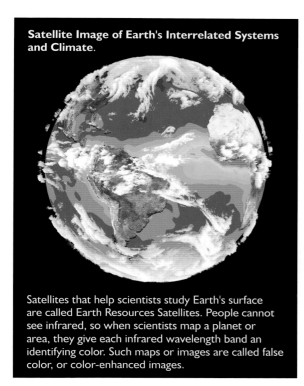

Satellite Image of Earth's Interrelated Systems and Climate.

Satellites that help scientists study Earth's surface are called Earth Resources Satellites. People cannot see infrared, so when scientists map a planet or area, they give each infrared wavelength band an identifying color. Such maps or images are called false color, or color-enhanced images.

Mount Etna. Plume of steam or ash from Mount Etna (2000).

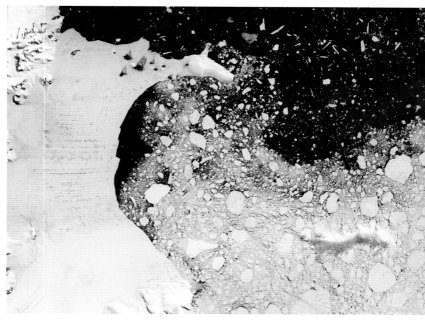

Antarctic Ice Seen from Space. View of icebergs splitting from the Larsen Ice Shelf in 2000. More than a tenth of Earth's surface is covered in ice, mostly in the ice caps at the Poles. The ice caps grow in the winter and shrink in the summer, when giant icebergs break off into the surrounding ocean.

of communities hit by floods or hurricanes help disaster relief efforts. And fast-food companies assess suburb growth to locate new restaurants.

Explorer XVII was the first satellite designed for the specific purpose of probing the Earth's atmosphere. Launched in 1963 into an egg-shaped orbit, which carried it through the layer of oxygen atoms, and into the helium belt once every ninety-six minutes, it repeatedly measured the composition, temperature, density and pressure of the gases that envelop the Earth.

Mercury astronaut Gordon Cooper demonstrated the potential of remote sensing early in the Space Program with his observations in 1963. During visual observations from the tiny porthole of his *Faith 7* spacecraft, he was able to see a test 44,000-watt Xenon lamp beamed at him from South Africa, as well as cities, oil refineries and even smoke from individual houses around the planet.

One of the first satellites designed for ocean surveillance was *ERS-1*, which carried instruments to measure infrared and microwave radiation.

Remote sensing spacecraft were developed from early meteorology satellites, which had simple onboard infrared cameras to image cloud formations by night. The first dedicated remote sensing satellite was *Landsat 1*, launched in 1972 equipped with a multi-spectral camera that transmitted digital data. There have been Landsat satellites in orbit ever since.

View of Lights at Night Around the World. Seen from Space, Earth is the only planet with strong signs of life. The evidence ranges from plants that change with the seasons and oxygen in the atmosphere to artificial radio signals and lights at night. This view identifies the major population centers on Earth.

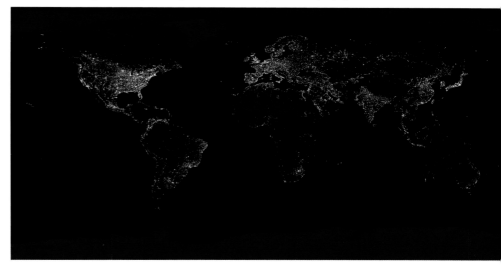

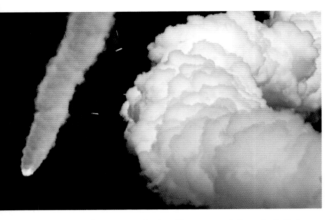

THEMIS Spacecraft on Its Way to Orbit. Right on schedule, six-solid rocket boosters fall away from the Delta II rocket carrying NASA's THEMIS spacecraft to orbit. The first six boosters were jettisoned after 66 seconds of flight. The rocket has nine boosters. This launch was on February 17, 2007 from Pad 17B at The Cape. THEMIS, or *Time History of Events and Macroscale Interactions during Substorms*, consists of five identical probes to track violent, colorful eruptions near the North Pole.

Landsat 1 took the first combined visible and infrared of Earth's surface. *Landsat 5* is still in service. *Landsat 7* was launched in April 1999. The Landsat satellites photograph almost the whole Earth every sixteen days.

In April 1986, when a nuclear reactor exploded in Chernobyl, Ukraine, a Landsat satellite passed overhead shortly afterward. Within seventy-two hours, a digital image of radioactivity pouring out of the reactor had been broadcast all over the world.

In 1978, the Seasat satellite made the first valuable measurement of oceans with radar.

NASA began the Earth Science Enterprise Program in 1991 to study Earth much more closely than it had. Cameras and other instruments onboard spacecraft in Earth orbit provide close-up images and other information about the planet's land, water, and air. These images and information are used in making plans to control damage from disasters, such as floods and wildfires, and to manage resources, such as farmlands and forests.

In September 1992, the *Topex/Poseidon* mission began collecting ocean data in unprecedented detail. From its 1,330-km-high orbit, the *Topex/Poseidon* satellite collected more data in a month than all research ships had in the previous hundred years.

In April 1994, images taken by a Space Shuttle *Endeavour* mission discovered the lost city of Ubar in the Sahara Desert in Africa. Instruments aboard the shuttle recorded images of a network of ancient tracks that helped pinpoint the city, which lay buried beneath the desert sand. Ubar, a remote desert trading outpost, vanished from recorded history about the year 300BC. Images taken from a Space Shuttle also revealed a hidden section of the ancient city of Angkor in Cambodia. During the ninth century, the city was home to more than 1,000,000 people. Today, a thick jungle covers much of it. Space Shuttle images detected canals north of the main city that cannot be seen from the ground.

At the beginning of the twenty-first century, there were about a half-dozen functioning X-ray satellites in orbit around the Earth. Two of these were NASA's *Rossi XTE* and *Chandra* satellites. Chandra, the X-ray equivalent of the *Hubble Space Telescope*, was launched in July 1999 and studies all X-ray sources in extraordinary detail.

Tropical rain forces are cleared every year to make way for land upon which farm animals can feed. In 2006, satellites showed that an area of Brazilian rain forest the size of Greece was destroyed in this way.

Meteorology Satellites

The way weather systems develop and move around the globe can be seen by meteorology satellites. They record images that are then broadcast nightly on television, show cloud cover, and monitor hurricanes growing and moving across the oceans. Meteorology satellites also carry instruments to take readings, which are converted to the temperatures, pressures, and humidity needed for weather forecasting. These, together with information from sources such as weather buoys, balloons, and ships, help forecasters improve their predictions.

The NASA space flights have provided many spectacular photographs of our planet, and among the most beautiful are those showing the swirling masses of cloud that trace the pattern of weather systems across the face of the globe.

TIROS

The *TIROS 1* (Television and Infra-Red Observation Satellite) was boosted into orbit by a Thor-Able rocket at Cape Canaveral on April 1, 1960. It recorded 22,952 cloud images from its orbit, including the first images from space of clouds moving. The *TIROS 1* spotted potential hurricanes far out at sea days before they would have been detected by any other means. It observed the spring breakup of ice in the St. Lawrence River and provided weather forecast for the re-supplying of Antarctic bases. *TIROS 2* launched later in the year,

TIROS Meteorological Satellite. TIROS (*Television and Infra-Red Observation Satellite*), the world's first weather satellite, was designed and built by RCA for NASA. The first TIROS was launched in 1960. Eight TIROS have been orbited successfully fulfilling their mission. Over 325,000 TV images were televised to Earth by TIROS. The TIROS, and other weather satellites, proved the value of weather watching from Space.

gave wider coverage and had infrared sensors so that it could send back pictures at nighttime as well.

In September 1961, the *TIROS 3* satellite sent TV pictures of a storm brewing in the Atlantic Ocean. Within days the biggest evacuation in U.S. history had saved 350,000 Gulf Coast inhabitants from Hurricane Carla.

The TIROS family of satellites improved with each successive launch. Later TIROS satellites added to their many accomplishments such feats as forecasting the weather for Mercury spacecraft launchings. Between 1960 and 1965, ten TIROS satellites aimed television cameras at Earth and sent back almost 650,000 images of worldwide weather systems. The last of the TIROS satellites, *TIROS 10*, was launched July 2, 1965, and shut down July 3, 1967.

Nimbus

NASA launched the *Nimbus 1*, the first polar weather satellite, on August 28, 1964.

In April 1970, it launched the *Nimbus 4* satellite, which carried the first instrument for measuring temperature at different altitudes in the atmosphere.

In 1973, the *Nimbus 7* weather satellite began keeping track of a portion of our atmosphere called the ozone layer. Produced through the interaction of solar energy and oxygen, ozone protects all living things from the Sun's damaging ultraviolet (UV) rays. Images provided by *Nimbus 7* showed that a huge hole had formed in the ozone layer. This hole, which appears over Antarctica for several months every year, is getting bigger. In 2000, the hole measured about eleven million square miles, larger than the entire North American continent. The images provided by NASA weather satellites helped convince countries to ban the use of certain chemicals that were determined to have caused the hole.

The Nimbus family of satellites, with its moth-like solar wings, served as a flying test bed for new instruments, which were later incorporated into several weather satellites.

Nimbus Weather Satellite. In April 1970, NASA launched the *Nimbus 4* weather satellite, which carried the first instrument for measuring temperature at different altitudes in the atmosphere.

GOES

The first Geostationary Operational Environmental Satellite (GOES) was placed in orbit in 1975, followed by a second one in 1977; eleven more GOES spacecraft have been launched since then. *GOES-13* was launched at Cape Canaveral by a Delta IV rocket on May 24, 2006.

GOES is a joint weather satellite program involving NASA and the U.S. National Oceanographic and Atmospheric Administration (NOAA), which owns and operates the satellites. NASA is responsible for spacecraft and instrument procurement, design and development, and its launch.

Today, two *GOES* satellites, positioned over the equator, give 24-hour coverage of the United States and South America. Each *GOES* satellite carries a visible-light imager and infrared sensor, plus instruments to measure X-rays and charged particles coming from the Sun.

Earth's Atmosphere. The Earth is surrounded by a thin layer of gas called the atmosphere, which protects its surface from the harshness of space. Heated unevenly by the Sun and spun around the Earth, the air is forced into ever-changing swirling patterns. The atmosphere is an essential blanket for life on Earth, keeping the planet at a comfortable temperature and protecting the surface from dangerous radiation. The atmosphere is a mixture of gases (mainly nitrogen and oxygen), water, and dust.

Hurricane Ivan in 2004. Geostationary satellites scan the region beneath them every thirty minutes. If a tropical storm develops, they can scan that region in more detail every fifteen minutes. The satellites also measure temperature, which helps forecasters predict hurricane strength. It is very difficult to predict the track of a hurricane, but due to the contributions of satellites, the accuracy of forecasts has improved by 0.5 to 1 percent each year during the past thirty years.

In the late 1990s, a joint effort between the United States and France known as Topex/Poseidon delivered images that have given scientists clues about El Nino, a phenomenon that periodically causes extensive flooding and extreme weather patterns. Topex/Poseidon shows the changes in ocean-wave height and average sea level that accompany El Nino. The satellite's reports are currently helping scientists predict El Nino events at least a year in advance.

Today, more than 120 countries from around the world receive most or all of their weather pictures from U. S. Satellites.

Military Satellites

Delta Vehicle-STSS-Delta Satellite Launch. Fire erupts across Launch Pad 17B at the Cape Canaveral as the United Launch Alliance Delta II rocket lifts off with the *Space Tracking and Surveillance System-Demonstrator* (STSS-Demo) spacecraft. The STSS-Demo was launched September 25, 2009 by NASA for the U.S. Missile Defense Agency. The STSS-Demo was a space-based sensor component of a layered Ballistic Missile Defense System designed for the overall mission of detecting, tracking and discriminating ballistic missiles.

Many of the earliest satellites were made for the armed services. Military satellites are widely used today. From the safety of orbit, satellites can gather information about battlefields, take pictures so detailed they can show where a person is standing, locate missing troops, and provide secure communications. Some satellites monitor the globe, watching for signs of the launch of a nuclear missile or a nuclear explosion.

Spy satellites, or photoreconnaissance satellites, usually fly in an orbit one hundred miles or more above the Earth's surface at speeds over 17,000-miles per hour

and produce high-quality, detailed images of battlefields, troop movements, missile sites, and other military targets.

The first American spy satellites were Discover satellites, launched in 1959. One aim of the Discover series was to perfect the technique of recovering an instrumented capsule from orbit. This was accomplished with *Discover 13* in August 1960, making it the first object to be retrieved successfully from space. The purpose of recovering such capsules was to obtain the high-resolution photographs they contained of ground installations such as missile sites and foreign air bases.

Another type of spy satellites, called *Samos* (Satellite and Missile Observation System), was developed and transmitted its pictures by radio from orbit. In 1971 a satellite called *Big Bird* combined the role of sending back pictures by radio and also, from time to time, ejecting film canisters. The *Midas* (Missile Defense Alarm System) satellite series carried infrared detectors to sense the hot gases of a rocket's exhaust. Other military satellites, most of which are classified and their names were not published, perform early-warning information, eavesdropping on enemy radio communications, enemy tracking functions and nuclear explosion detection. Hundreds of spy satellites, from many different countries, now orbit the Earth.

Missile early warning is the mission of the United States Military Defense Support Program (DSP) that provides worldwide coverage by using a network of satellites in geosynchronous orbit to monitor the Earth for Space and missile launches. The first DSP satellite was launched in November 1970. The satellite uses infrared sensors to detect heat from missile plumes against the Earth's background. With up to ten satellites still operational, they can provide stereo views of launches and better plume characterization.

Most DSP satellites were launched into orbit by a Titan IV booster. *DSP-16* was launched from Space Shuttle *Discovery* on mission *STS-44* on November 24, 1991. *DSP-23* was sent into orbit by a Delta IV rocket on November 10, 2007.

The DSP was originally designed to detect strategic threats against the continental United States; however, it also aided the military in the Gulf War of 1991. The DSP satellites turned their attention to the Persian Gulf and detected Iraqi ballistic missiles being launched against the U.S.-led coalition forces.

Because of the DSP's global satellite coverage, no missile launch, no matter where in the world it came from, can go unnoticed. It is one of the U.S. military's most successful programs.

The military has started to replace the older DSP satellites with a fleet of satellites called the Space-Based Infra-Red System (SBIRS). These satellites carry sensors to detect missile launches.

AMBASSADORS TO THE SOLAR SYSTEM

THE SOLAR SYSTEM IS AN ORDERLY COMMUNITY OF EIGHT PLANETS AND their Moons, dwarf planets, and myriads of asteroids and other small bodies, many of them sweeping in regular orbits around the Sun, the center of the Solar System.

The Solar System is the region in which all our spacecraft, so far, have explored. While our earthly and space-born telescopes have probed to the depths of the Universe, our craft have gone only just beyond the edge of the planetary system.

THE SUN

At the center of the Solar System is the brilliant nuclear furnace we call the Sun. It is a huge globe of hot gas. It is 109 times the diameter of Earth and has a mass 1,000 times greater than that of all the planets in the Solar System put together. The Sun's nuclear furnace has been blazing for 4.6 billion years.

The Sun is composed of about seventy-five percent hydrogen, twenty-five percent helium, and a trace of other elements. The Sun's energy is derived from the conversion of hydrogen to helium.

The Sun may seem very special to us, but in Space terms it is nothing more than a very ordinary, medium-sized star that make up our local galaxy — the Milky Way — and beyond this, there are millions of other galaxies in the Universe, some of them even larger than the Milky Way. Our Sun and its cluster of planets lie about two-thirds of the way out from the center of the Milky Way, on one of the long spiral arms.

Distances in Space are so great that they are measured in light-years. A light-year is the distance traveled by a ray of light in one year, roughly six trillion miles. The Milky Way is about 100,000 light-years across.

Several NASA missions have collected information about the Sun — some of them while they were on their way to do something else. In the 1960s, several of the *Pioneer* spacecraft voyages returned information about the Sun. In the 1970s, astronauts aboard the *Skylab Space Station* observed the Sun, and the *Helios* spacecraft provided close-up studies of cosmic rays and solar dynamics. In 1990, the spacecraft *Ulysses* was launched from the Space Shuttle *Discovery*. It went into orbit over the Sun's poles, studied solar wind, and examined the nature of the Sun's corona and magnetic field.

Seeking to understand the workings of the Sun, NASA and European Space Agency (ESA) scientists launched the *Solar and Heliospheric Observatory* (SOHO) into solar orbit December 2, 1995. The SOHO spacecraft carried an array of twelve instruments that continue to map, record, analyze, and monitor solar activity, enormously increasing our knowledge of the Sun. The SOHO spacecraft has recorded much of the Sun's furious activity, especially during the years 2000-2003.

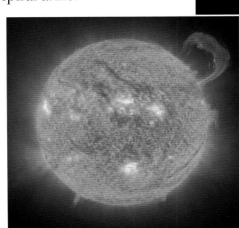

The Sun with a Handle-Shaped Prominence. Prominence is a mass of gas hanging in the Sun's atmosphere. Huge clouds and sheets of gas, or prominences, extend upward from the chromosphere (a dense layer of hydrogen and helium gas), stretching hundreds of thousands of miles into the corona (the outermost region of the Sun's atmosphere). Some solar prominences can persist for weeks or months. This photo was taken by the SOHO spacecraft in 2003.

Solar Eruption on the Sun. In addition to giving off light and heat, the Sun, during solar flares, billows forth clouds of particles and other emissions of varying intensity. When these particles are strong enough and collide with the Earth some time later, they can dissipate more energy in the Earth's high atmosphere than the most destructive hurricanes on record. The resulting electrical and magnetic disturbances can black out long-range radio communications, cause airplanes and ship compasses to shift erratically and trigger brilliant aurora flashes.

On January 4, 2008, SOHO recorded an event solar scientists had been eagerly anticipating — the appearance of a reverse-polarity sunspot (magnetic storm) high above the latitude of the Sun's equator.

Also in 2008, SOHO confirmed a thirty-year-old theory that massive solar flares can set the entire surface of the Sun oscillating, rippling in waves that scientists have dubbed a "starquake." SOHO also discovered that the Sun is home to massive tornadoes that form funnels spinning at 310,685 mph.

Scientists are expecting SOHO to produce many discoveries in 2012 when they think the next furious burst of activity on the Sun will occur.

from a few feet to several hundred miles across. Farther out are the four gas giants—Jupiter, Saturn, Uranus, and Neptune—which are huge frozen worlds of hydrogen, helium, and other gases. And beyond them lies Pluto, a tiny ball of dust and ice spinning through Space like a dirty snowball, almost four billion miles from the Sun.

All of the planets in the Solar System travel around the Sun in the same direction, on paths or orbits, that are oval in shape. From the Sun outward, the order is Mercury, Venus, Earth, Mars, asteroids, Jupiter, Saturn, Uranus, Neptune, and the dwarf planet Pluto.

PLANETS IN THE SOLAR SYSTEM

The four innermost planets in the Solar System are solid, but only Earth has water on its surface and a life-supporting atmosphere. In late 2009, NASA discovered water on Earth's Moon. Mercury and Venus are closer to the Sun and too hot for life. Mars is farther away and too cold.

Beyond these inner planets lies the asteroid belt, a zone of rocks and mini-planets ranging in size

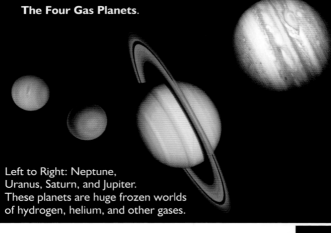

The Four Gas Planets.

Left to Right: Neptune, Uranus, Saturn, and Jupiter. These planets are huge frozen worlds of hydrogen, helium, and other gases.

The Eight Largest Planets. Top to Bottom: Mercury, Venus, Earth with Moon, Mars, Jupiter, Saturn, Uranus, and Neptune.

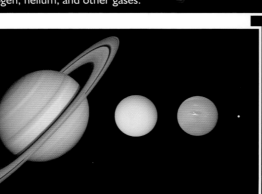

Planets in the Solar System. Left to Right (distance from the Sun): Mercury, Venus, Earth, Mars, Jupiter, Saturn, Uranus, Neptune, and the dwarf planet Pluto. The Sun is located on the left.

✦ **Mercury** is about 3,025 miles in diameter; it has no Moons, and no atmosphere.

✦ **Venus** is 7,500 miles in diameter. It has no Moons and a dense atmosphere of carbon dioxide gas.

✦ **Earth** is 7,910 miles in diameter. It has one Moon and an atmosphere made mainly of nitrogen and oxygen.

✦ **Mars** is about 4,220 miles across. It has two Moons and a thin atmosphere made of carbon dioxide.

✦ **Jupiter** is 88,500 miles across. It has sixteen or more Moons and one faint ring.

✦ **Saturn** is 74,90 miles in diameter. It has many Moons and several bright rings.

✦ **Uranus** is 32,240 miles in diameter. It has vertical rings and seventeen faint Moons.

✦ **Neptune** is 30,758 miles in diameter. It has thirteen Moons and several rings.

✦ **Dwarf Planet Pluto** is about 1,365 miles across. It has three small Moons. Very little is known about this planet.

Probes to the Planets

Once the first rockets had successfully reached space, a new investigative tool, the Probe, became available to scientists and astronomers. Probes are car-size robot craft launched by rockets. They travel to a predetermined target and investigate it using their onboard instruments. Probes have given us close-up views of the Sun, planets, Moons, a comet, and asteroids. They have taught us much about the Solar System.

A probe may fly by a target, orbit it, or land on it. A flyby probe surveys its target as it flies past. For example, the flyby probes *Pioneer 10* and *Pioneer 11* left Earth in 1972 and 1973 respectively, and headed for the outer planets. *Pioneer 10* was the first spacecraft to cross the asteroid belt, the region between Mars and Jupiter containing a large concentration of asteroids. In 1973 *Pioneer 10* reached Jupiter and took the first close-up photographs of the giant planet. It then kept traveling, and left the Solar System in 1983. *Pioneer 10's* signals faded away in 2003 when it was 7.6 billion miles from Earth on its way toward the star Aldebaran.

The *Pioneer 11* also headed first to Jupiter, using that planet's gravitational field to propel it toward Saturn. It arrived at Saturn in 1979 and proceeded to photograph and collect other valuable information about its rings and moons. In 1990, *Pioneer 11* exited the Solar System and, in September 1995, after twenty-two years of operation, its power supply ran out.

The *Pioneer 10* and *11* missions showed scientists that spacecraft could safely navigate the debris-strewn route to the outer planets.

The final two probes bearing the name Pioneer were launched in 1978 to explore Venus. The first, called *Pioneer-Venus 1*, studied the planet's atmosphere, mapped about ninety percent of its surface, then descended to the planet in 1992, and burned up. The second probe, *Pioneer-Venus 2*, was launched three months after the Orbiter. The *Pioneer-Venus 2* distributed five probes around the planet, one of which transmitted data from the surface for sixty-seven minutes.

The flyby probe *Voyager 2* investigated Jupiter, Saturn, Uranus and Neptune between 1979 and 1989.

Probes have landed on the Moon and Venus. Several probes and rovers have landed on Mars. They have studied, orbited, mapped, photographed, and probed. They have scooped up soil, run experiments, and even trundled across the surface in some places. They have trained their instruments on the Sun.

MERCURY

Mercury is the planet closest to the Sun, and is the smallest of the planets. It revolves around the Sun every eighty-eight days. It has some of the most extreme temperatures in the Solar System, with a maximum surface temperature of 800°F and a minimum of minus 280°F. This dry, rocky world has an atmosphere so thin that it barely exists. Of all the planets in the Solar System, Mercury travels around the Sun the fastest but spins slowly on its axis.

Mercury, First Planet from the Sun. The innermost planet, Mercury, was viewed by the *Mariner 10* spacecraft in 1974. The *Mariner 10's* cameras took sequences of overlapping photographs as the spacecraft passed high above the surface. The pictures were radioed back to Earth, where they were enhanced by computers and then combined to give an overall view of the planet. Craters cover about sixty percent of Mercury's known surface. The rest is mostly lava plains.

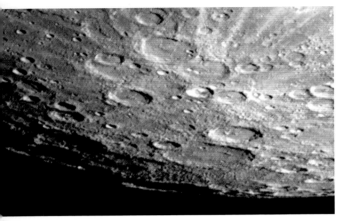

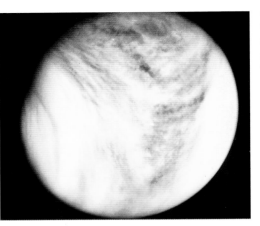

Mercury's South Pole Region. Mercury's polar regions include areas that are always shaded from the Sun's heat, in which ice may exist. Mercury's rugged surface is like the Moon's and crowded with impact craters.

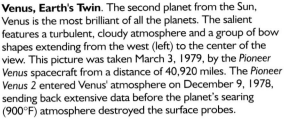

Venus, Earth's Twin. The second planet from the Sun, Venus is the most brilliant of all the planets. The salient features a turbulent, cloudy atmosphere and a group of bow shapes extending from the west (left) to the center of the view. This picture was taken March 3, 1979, by the *Pioneer Venus* spacecraft from a distance of 40,920 miles. The *Pioneer Venus 2* entered Venus' atmosphere on December 9, 1978, sending back extensive data before the planet's searing (900°F) atmosphere destroyed the surface probes.

Computer Enhanced View of Venus. The topography of Venus stands out sharply in this computer-produced global view based on radar data from the *Pioneer Venus* spacecraft. To accentuate the planet's relief, the artificial color scheme shows the lowest-lying areas in blue, moderate elevations in green, higher elevations in yellow, and the highest in red. The apparent hole at the top is an area from which no data was received.

Cloud-top View of Venus. From Earth we can see only the cloud tops of Venus. Hidden under this thick blanket of gas is a landscape molded by volcanic eruption. The Sun's heat drives the clouds around Venus. The clouds move in the same direction as the spinning planet, but sixty times faster.

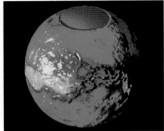

About four billion years ago, in the early history of the Solar System, young Mercury's surface was punctured by meteorite impacts. Lava flooded out from the interior to form extensive plains, giving the planet an appearance that, at first glance, resembles the Moon. With no wind or water to shape its crater-scarred landscape, Mercury has remained virtually unchanged since then. It is composed mostly of iron.

From Earth, faint markings can be seen on the planet's surface, but our only close-up views date from the 1970s, when the planetary probe *Mariner 10* flew by and revealed Mercury to be a heavily cratered world.

On March 29, 1974, *Mariner 10* gave us our first view of the Moon-like of Mercury, from a distance of only 168 miles. The probe, in solar orbit, passed Mercury two more times (September 1974 and March 1975), sending back more than 12,000 pictures, together with magnetic and radiation measurements. The pictures showed that Mercury's surface is parched and pitted; it's also pockmarked by thousands of craters caused by meteorite impacts. Mercury's craters covered almost the entire area photographed by *Mariner 10*, which also determined that the planet has a magnetic field.

In January 2008, the *Messenger* probe flew through Mercury's magnetic field and detected charged particles of silicon, sodium, sulfur, and even water icons, blasted from the planet's surface and atmosphere by the solar wind. The *Messenger* probe flyby provided a part of the planet never before viewed by spacecraft and returned a gold mine of new data.

Two more probes are set to circle the planet in the near future.

VENUS

Venus is a sphere of rock similar in size to Earth, however, it is a dark hostile world of volcanoes and suffocating atmosphere. It is closer to the Sun than Earth and its temperature is higher than that of any other planet. Venus' dense atmosphere captures so much of the Sun's heat that the mean surface temperature is about 850°F.

The Venus atmosphere is mainly carbon dioxide but also includes sulfuric acid from the planet's many volcanic eruptions. This hostile atmosphere makes Venus a hot, gloomy, suffocating world.

Venus is the second nearest planet to the Sun. It spins backward very slowly, taking 243 days to complete one rotation, but it has the most circular orbit of all the planets.

Venus has been studied by five major NASA missions — *Mariner 2* flyby in 1962, *Mariner 5* in 1967, *Mariner 10* (on its way to Mercury) in 1973, *Pioneer-Venus 1* (orbiter) and *Pioneer-Venus 2* (five atmospheric probes) in December 1978, and *Magellan* (orbiter) in 1990-1994.

On June 19, 1963, the *Mariner 2* spacecraft passed within 22,000 miles of Venus. It showed that Venus was a forbidden planet. The next American Venus probe was *Mariner 10*, launched in 1973, which flew past Mercury too. The first two-planet explorer passed within 3,580 miles of Venus on February 5, 1974, transmitting the first pictures of the planet's lethal cloud cover. Venusian gravity then bent the probe's trajectory for its rendezvous with Mercury.

Pioneer-Venus 1 used radar to create a detailed map of the surface of Venus. The radar images showed objects as small as sixty miles across and provided scientists with information about ninety percent of the planets surface. For the first time, humans could peer through Venus' thick, cloudy atmosphere and see the surface below. The map showed continent-like highlands, hilly plains, mountains, and flat, barren lowlands. *Pioneer-Venus 2* probes helped scientists understand why Venus is hotter than Mercury.

During four 243-day-long mapping sessions between 1990 and 1994, the *Magellan* probe gathered data covering ninety-eight percent of the Venusian surface. Two radar beams surveyed long, narrow strips of land below the probe's flight path. Data from the strips was combined to produce images of large areas of Venus.

Mars

The planet Mars was named after the Roman god of war. Sometimes known as the Red Planet, it is composed of dense, rocky material, and along with Mercury, Venus, and Earth, it is one of the four terrestrial, or Earthlike, planets of the inner Solar System. Mars is one and a half times more distant than Earth from the Sun. In the late 1990s, scientists began to study Mars in unprecedented detail.

Mars has two Moons, Phobos and Deimos. They are among the darkest objects in the Solar System because they reflect very little light. Mars is a rocky planet with an iron-rich core. Much of the planet's surface is a frozen rock-strewn desert interrupted by dunes and craters, however Mars also has volcanoes and canyons that dwarf those found on Earth. The planet's red color comes from soil rich in iron oxide (rust).

Mars is about half the size of Earth, but is less dense and weighs only about one-tenth as much. The Martian surface covers about as much area as the continents on Earth. The Martian atmosphere is about ninety-five percent carbon dioxide and three percent nitrogen. Earth's atmosphere is seventy-eight percent nitrogen and twenty-one percent oxygen.

The daytime high temperature at the surface of Mars is 65°F, but the temperature drops dramatically only a few feet above the surface. The daytime high five feet above the surface is only 15°F. At night, temperatures drip to a low of minus 130°F at the surface and minus 105°F at five feet.

Mars has seasons similar to those on Earth. Its elliptical orbit, in contrast to Earth's nearly circular orbit, means its distance from the Sun also affects its seasons.

The *Mariner 4* spacecraft, on July 14, 1965, passed within 6,000 miles of Mars and sent back twenty-two pictures of craters gored into an arid landscape. Scientists thought it resembled the Earth's Moon.

Four years later, in 1969, the next two Mars probes, *Mariner 6* and 7, sent back 201 pictures from within 2,200 miles of Mars. *Mariner 8* crashed into the Atlantic Ocean after its rocket broke up.

The *Mariner 9* became the first man-made object to orbit another planet on November 14, 1970. On January 7, 1971, cameras aboard *Mariner 9* photographed a staggering view of Olympus Mons, a huge volcano broader than any on Earth. It was fifteen miles high — almost three times the height of Everest on a planet half the size of Earth. *Mariner 9* mapped the entire surface of Mars, took 7,329 photographs and sent back fifty-four billion items of scientific information. These photographs also revealed the canyons of Valles Marineris on the planet, a vast canyon system with tributaries as large as Earth's Grand Canyon.

Viking I Launch from Cape Canaveral. Atop a Titan-Centaur rocket, the *Viking I* spacecraft, containing the first Mars lander, is blasted off from Cape Canaveral August 20, 1975. In July 1976 the spacecraft arrived at Mars and released a lander that parachuted to the surface carrying sophisticated experiments designed to pick up the telltale signs of living organisms in the Martian soil.

On June 9, 1976, *Viking 1* made the longest deep-space engine burn in history to place itself in orbit around Mars. It took the best part of a year to complete its 440-million-mile journey. The *Viking 1* released a three-legged, six-foot-tall robot on the Martian surface on July 20, 1976, the seventh anniversary of the first manned Moon landing. Immediately after touchdown, one of the robot's two cameras recorded the first black and white picture from the surface of Mars. Six minutes later it began a 300-degree panorama of the Mars-scape. The pictures were astonishingly good. Later color pictures were even more remarkable, proving at last that the surface of Mars really is red.

The second Viking lander, *Viking 2*, touched down on Utopia Planitia on September 3, 1976, and began photographing Mars' surface. The *Viking 2* lander even sent pictures of ground frost on a chilly —114°F — winter morning. Both Viking landers carried sophisticated experiments designed to pick up telltale signs of living organisms in the Martian soil. Although the tests showed no evidence of life on Mars, the question still appears to be open.

Mars Global Surveyor was launched in 1996. It has sent back information that leads scientists to think a great deal of frozen water may still exist below the planet's surface, and where there is water, life may have existed at one time. The *Mars Global Surveyor* carried a high-resolution camera and instruments that allows it to measure magnetic fields, temperature variations,

and the altitudes of surface features. It was designed to map the planet from polar orbit and to search for signs of water and geological activity. The spacecraft's mission ended November 2, 2006.

NASA's lightweight *Pathfinder* spacecraft landed on the Martian surface July 4, 1997.

Two days after the *Pathfinder* touched down on the Martian surface, the six-wheel Surface Rover, *Sojourne*, rolled out of the lander onto the planet's surface to explore its surroundings. Between July 4 and September 27, when communication was lost, *Pathfinder* transmitted a phenomenal amount of information that included 16,550 images of the environment along with extensive data concerning Martian winds, weather and rocks.

In 1977, the *Mars Global Surveyor* mission reached Mars. The U.S. was stung by two major failures in the year 1999, when two expensive missions, the *Mars Polar Lander* and the *Mars Climate Orbiter*, were lost.

The *Mars Odyssey*, launched April 7, 2001, was designed to search for water, volcanic activity and minerals, and to serve as advanced communications link to send data from rovers. The spacecraft's mission is slated to end in September 2010.

The *Mars Express Orbiter*, launched June 2, 2003, searched for water and signs of past or present life. It carried a small lander, the *Beagle 2*, which was released December 19, 2003, but failed to maintain contact. The *Express Orbiter's* mission ended December 31, 2009.

The *Spirit Rover*, launched June 10, 2003, and the *Opportunity Rover*, launched July 8, 2003, were designed to study rocks, soil, and minerals on Mars' surface; to explore geological processes; and to search for past and present signs of water. The rovers are still operating.

The *Mars Reconnaissance Orbiter*, launched August 12, 2005, was designed to search for water on Mars; it carries powerful cameras and has

LEFT COLUMN:
Spacecraft Views of Mars Surface.
1. Sulfur-rich rocks and surface materials on Mars; 2. Dune fields on floor of Endurance Crater on Mars; 3. Basaltic or volcanic rocks on Mars; 4. Magnified look at soil on Mars.

NEAR LEFT:
Dunes and Rocks on Mars. Colorized view of Mars form the Mars exploration rover *Spirit* in 2004, showed beautiful dunes and scattered rocks. Distribution of rocks on the surface indicates that large bodies of water once flowed on Mars.

advanced telecommunications that allow it to send messages to and from rovers on surface. The Hi-RISE (High Resolution Imaging Science Experiment) camera has sent the most detailed pictures of Mars yet taken. The Orbiter mission is slated to end on December 31, 2010.

The *Mars Phoenix Lander*, launched August 4, 2007, touched down near Mar's North Pole to study ice deposits beneath the surface with a robot arm and scoop. On July 31, 2008, the *Phoenix* made history by confirming the presence of water ice on Mars. It went out of service when the planet's winter season halted its solar power. The Phoenix mission ended November 2, 2008.

JUPITER

The fifth planet from the Sun is very different from Earth-like planets. Jupiter is the largest planet in the solar system — more than three hundred Earths would fit into its volume and its mass is larger than all the other planets combined. Jupiter is a gas giant, mostly hydrogen and helium); unlike the small, rocky inner planets, it has no solid surface, but is all gas and liquid except for a very small rocky core. The planet has sixteen or more Moons and a dusty ring system. Jupiter's four principal Moons are Callisto, Europa, Ganymede, and Io. The other twelve Moons are Metis, Adrastea, Amalthea, Thebe, Leda, Himalia, Lysthea, Elara, Ananke, Carme, Pasiphae, and Sinope.

On March 2, 1972, *Pioneer 10* left Earth faster than any man-made object had traveled before, covering nine miles every second as it headed toward Jupiter. The 595 pound probe carried eleven experiments powered by four small nuclear generators, and could transmit information to Earth from beyond 9,000-million-miles.

On December 2, 1973, the *Pioneer 10* spacecraft passed 81,000 miles above the cloud tops, snapping pictures and measuring the environment of Jupiter and its Moons. In 1987, *Pioneer 10* became the first spacecraft to leave the Solar System, and it should reach the vicinity of star Aldebaran about the year 8,002,000 A.D.! Should some intelligent being from another star system come across a wandering *Pioneer 10* spacecraft they will find an engraved plaque depicting a man and a woman of our species as they looked in the days when humans first left the cradle Earth. *Pioneer 10* discovered Jupiter's unusually massive magnetic field. Jupiter's magnetic field is 20,000 times stronger than Earth's. This field reaches out into Space, surrounding the planet in a huge magnetic bubble, or magnetosphere.

The *Pioneer 11* spacecraft repeated *Pioneer 10's* success a year later passing 26,000 miles from Jupiter on December 3, 1974. Jupiter's gravity catapulted the spacecraft back across the Solar System for a rendezvous with Saturn.

Voyager 1, launched on September 5, 1977 and *Voyager 2*, launched on August 20, 1977, were identical spacecraft. They took separate paths on their way to Jupiter. After a 500-million-mile journey, *Voyager 1* arrived at Jupiter in early March 1979, followed four months later by *Voyager 2*. A slingshot trajectory sent both probes toward Saturn. Both of the Voyager spacecraft carry messages from Earth for intelligent beings should they find them floating around in Space millions of years from now.

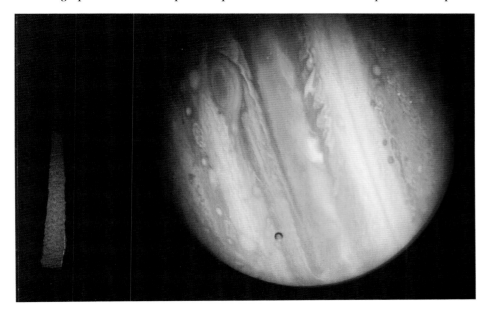

Jupiter's Great Red Spot. The Great Red Spot is a hurricane-like weather formation, with an anticyclonic direction of circulation. It is somewhat higher than the surrounding atmosphere; its color comes from red phosphorus, a product of the decomposition of phosphine. Jupiter's Great Red Spot formation is large enough to engulf Earth. This photograph, taken by *Voyager 1* in 1979, reminds one of modern art.

Turbulent, Stormy Jupiter. The atmosphere of Jupiter, fairly turbulent and brightly colored, is divided into light zones and dark belts. Clouds of icy ammonia crystals form the lighter zones. The dark belts generally indicate lower-lying atmospheric regions made up of hydrogen and compounds like ammonium hydrosulfide. The picture was taken by *Voyager 1* on February 5, 1979, from a distance of 17.5 million miles. The planet's Great Red Spot (top, left side) has intrigued astronomers for more than a century.

A Jupiter ring system was observed by *Voyager 2*. It consists of extremely fine particles of dust about twenty miles wide. The dust particles extend in toward Jupiter to the top of the atmospheric clouds. The outer edge of the ring is hard. On Jupiter's Moon, Io, eight volcanoes erupted while *Voyager* was snapping pictures. This was the first ongoing volcanic activity anyone knew of outside of the Earth. *Voyager* also detected a strange, tube-like flow of electricity between Jupiter and Io.

Most spectacular, though, were the Voyager's images of the big Moons that hover close to Jupiter. Callisto, Europa, Ganymede, and Io are much more fascinating than scientists had imagined. Callisto has a lifeless, ancient surface while Io is hot and volcanic, spewing sulfur and seething with activity. Ganymede is full of mystery, and Europa seems icy and strange.

In December 1995, NASA's *Galileo* spacecraft arrived at Jupiter. Its orbiter and probe have revolutionized knowledge of the planet. *Galileo's* orbiter circled the planet ten times in twenty-two months while a separate probe descended into the atmosphere. *Galileo* returned stunning and intriguing results in 1998 and 1999. One of the biggest surprises was evidence that the Moon, Europa, may have a liquid ocean, and possibly, some form of life beneath its icy exterior. The *Galileo* orbiter made numerous close-up pictures of the rugged, cratered surface of the Moon, Callisto.

The *Cassini* space probe flew past Jupiter en route for Saturn in 2000.

The *Hubble Space Telescope* and supporting observations of two giant storms erupting on Jupiter in 2007 showed that such tempests are triggered deep down in water

Cassini Photograph of Jupiter. The *Cassini* space probe flew past Jupiter en route for Saturn in 2000. The black spot, lower left corner, is the shadow of the Moon Io.

Callisto, Jupiter's Battered Moon. Consisting of sixty percent rock and iron and forty percent ice and water, Callisto's surface is completely covered with craters and its concentric ridges, formed by a huge hurtling body, radiate from a bright core. Callisto is the Solar System's most heavily bombarded Moon, has a diameter of more than 2,980 miles, and also has the oldest terrain in the Solar System. Lying 1.17 million miles from Jupiter, Callisto makes one orbit in a leisurely 16.69 days. *Voyager 2* took this view July 7, 1979, from a distance of 1,438,000 miles.

Groovy Ganymede, Jupiter's Largest Moon. Ganymede is bigger than the planet Mercury. It has a diameter of 3,280 miles and thought to have a rocky core surrounded by water ice and a crust of ice and rock. Ganymede lies 664,900 miles from Jupiter and circles it once every seven days and three hours. The *Voyager 2* satellite photographed this view on July 2, 1979, from a distance of four million miles.

Cracking Europa, Jupiter's Fourth Largest Moon. With a diameter of 1,950 miles, Europa lies 417,000 miles from Jupiter, orbiting the planet every three days and fourteen hours. Water and ice may have oozed like lava into cracks and craters, making this the Solar System's smoothest Moon. Nevertheless, Europa looks like a badly cracked eggshell in this photograph. The surface of Europa is smooth ice and evidence from the *Galileo* probe points to the existence of a liquid ocean beneath the ice. The *Hubble Space Telescope* has also detected a thin atmosphere of oxygen on Europa. Taken on November 6, 1996, this photo shows the faintly mottled and streaked icy plains of Europa.

Explosive Io, Jupiter's Third Largest Moon. The most volcanically active body in the Solar System, many giant volcanoes erupt on Io at any given time. With a thin atmosphere of sulfur dioxide, Io is covered with volcanoes, lava flows, and high mountains. It has a diameter of 2,256 miles, orbits Jupiter in 1.77 days, and is 263,000 miles from Jupiter. This *Galileo* false color composite image shows two volcanic plumes on Io as they appeared on June 28, 1997. The bright colors are due to sulfur and sulfur compounds spewed from the many volcanoes, the dark spots peppering its surface. The white patches may be sulfur dioxide snow. Mountainous regions exist near both poles.

Moons of Jupiter. Jupiter has sixteen known Moons. The largest four are Io, Europa, Ganymede, and Callisto. Galileo first investigated these four Moons in 1610. Fittingly, it is a probe called *Galileo* that has revealed how the complex elements of the Jovian system work together and affect one another. The Galilean Moons orbit Jupiter in nearly circular paths almost exactly around the planet's equator.

clouds by hot jets that send ammonia ice racing at 375 miles per hour high above the visible cloud layer.

NASA is sending another spacecraft to orbit Jupiter and continue *Galileo's* fine work. The solar-powered *Juno* is scheduled to launch in 2011 on a five-year journey. After arriving in 2016, it will probe deep into the planet's atmosphere.

SATURN

Saturn — Sixth Planet from the Sun. Saturn's rotation period around the Sun is 10,759 Earth days or 29.46 Earth years. Described as the most beautiful planet, it is easily recognizable because of the bright rings around its equator. This photograph was taken by *Voyager 1* on October 5, 1980, from a distance of 32 million miles.

Like Jupiter, Saturn, the second largest planet in the Solar System, is a large ball of gas and liquid topped by clouds. It has many Moons and the most extensive ring system of any of the planets. One of its rings, the "F" ring, is made of two narrow bright rings with a fainter ring inside them, all of which combine to give it a braided appearance.

Voyager II Image. Saturn is famous for its system of rings, a distinction it shares to some extent with the other gas-giant planets. *Voyager 2* took this photograph, on April 8, 1981, through green, violet and ultra-violet filters, revealing the planets distinct bands in its dense, stormy atmosphere.

Saturn's Rings. The rings around Saturn are composed of silica rock, iron oxide and ice particles. The Saturn rings stretch farther than the rings of any other planet. Saturn's rings are probably the remains of one or more captured comets that have broken up, probably within the past few hundred million years. Space probes show that the particles in the main rings are arranged in thousands of closely packed ringlets. The *Voyager* spacecraft found that Saturn's ring system was composed of thousands of rings. The colors in this photograph are not real, but are added by computer to highlight the differences in the rings.

Phoebe — Saturn's Outermost Moon. The outermost Moon, Phoebe, orbits in the opposite direction from the other Moons. Phoebe orbits Saturn every 550.48 Earth days. Heavily cratered Phoebe appears to be an ice-rich body coated with a thin layer of dark material.

Exciting Titan, Saturn's Largest Moon. Titan is Saturn's largest Moon, or satellite, and, with a diameter of 3,200 miles, it is the second biggest Moon in the Solar System. It is mainly composed of water ice and rock and has a dense atmosphere. The *Cassini* probe's cloud-piercing radar took the shown picture of Titan's own lake district, replenished by raining fuel such as methane. The Huygens lander, which made a soft landing on Titan in January 2005, sent back a photograph of a rocky landscape.

Saturn and Six Moons. This montage of the Saturnian system is a composite of several images taken by *Voyager 1* during its Saturn encounter in November 1980. Shown are Dione in front, Saturn rising behind, Tethys and Mimas fading in the distance to the right, Enceladus and Rhea off Saturn's rings to the left, and Titan in its distant orbit at top. Titan, Saturn's largest Moon, is the second largest Moon in the Solar System. It is bigger than the planet Mercury.

Rhea with Saturn in Background. Rhea orbits Saturn every 4.52 Earth days. Saturn's Moon Rhea has a surface peppered with small craters. It is 956 miles wide and Saturn's second biggest Moon. The *Cassini* probe photographed possible fractures in the surface, and in March 2008, encircled by faint dusty rings.

Saturn has a diameter of 74,900 miles, is located 934,340,000 miles from the Sun, has a core of rock and metallic hydrogen, an atmosphere composed of ninety-four percent hydrogen and six percent helium, spins on its axis once every 10.5 hours, and orbits the Sun every 29.46 Earth years.

Eighteen of Saturn's Moons are Pan, Atlas, Prometheus, Pandora, Epimetheus, Janus, Mimas, Enceladus, Tethys, Telesto, Calypso, Dione, Helene, Rhea, Titan, Hyperion, Iapetus and Phoebe.

Four probes have flown past Saturn — *Pioneer 11*, *Voyager 1* and *2*, and *Cassini-Huygens*.

The day before *Voyager 1* swept by Saturn, it flew within 2,500 miles of Saturn's largest natural satellite, Titan, and passed behind it. As *Voyager 1* approached Saturn, it detected that it is visible flat at the poles due to rapid rotation. Radio emissions from the planet determined the length of Saturn's days to be ten hours, thirty-nine minutes, and twenty-four seconds. Winds blow at extremely high speeds on Saturn as *Voyager 1* measured winds near the equator of 1,100 miles per hour. The atmosphere is dense and consists mainly of hydrogen and helium. Saturn's clouds are mostly frozen crystals of ammonia ice.

Voyager 1 and *2* cameras photographed Saturn's rings and all of the natural satellites of Saturn. Saturn's satellites are composed of thirty to forty percent ice. Enceladus, for example, is almost pure water ice. Seventy thousand pictures later, the Voyager spacecrafts completed their photographic missions.

Voyager 2 traveled on to Uranus and Neptune.

Views of Saturn's icy Moon, Titan, and its hazy, orange atmosphere intrigued scientists so much that NASA launched the *Cassini-Huygens* spacecraft in September 1997 to take a closer look at Saturn and Titan. In March 2008 the spacecraft recorded a long-lived electrical storm on Saturn.

URANUS

Uranus is the third largest planet in the Solar System. It has seventeen known Moons and eleven known rings, which circle the planet's equator. Its atmosphere has a blue-green color produced by methane and high-altitude smog.

The composition of its outer layers is similar to that of Saturn and Jupiter. The visible part of the planet seems to be almost featureless. Uranus takes more than eighty-four years to orbit the Sun.

The Moons, or natural satellites, of Uranus are Cordelia, Ophelia, Bianca, Cressida, Desdemona, Juliet, Portia, Rosalind, Belinda, Puck, Miranda, Ariel, Umbriel, Titania, Oberon, Caliban, and Sycorax. The *Voyager 2* spacecraft discovered the first ten listed in 1986.

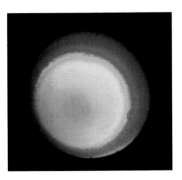

Uranus — Seventh Planet from the Sun. Uranus is a gaseous planet like Jupiter and Saturn, but with a distinct blue-green appearance. It rotates the Sun every eighty-four Earth years. This view has been color enhanced by a computer to bring out details in the polar region. Uranus has a diameter of 32,116 miles. This photo was taken by a 1,500-mm camera from the *Voyager 2* spacecraft, which took 5,800 pictures as it came within 51,000 miles of the planet's cloud tops on January 24, 1986.

Miranda, Uranus' Moon. The smallest of Uranus' five major Moons, Miranda's surface is jumbled with a bright check mark and grooves. One theory is that it broke apart and came together again. Its composition is half water-ice, and the balance is silicate rock and methane-related organic compounds. It has fault canyons twelve miles deep, cracks, lava flows, and strange tracks.

In January 1986, the *Voyager 2* spacecraft's images of Uranus revealed the planet's smooth, blue-green cloud tops and thin rings of Uranus with shepherd Moons much like the ones orbiting Saturn. The spacecraft also revealed ten previously undefined Moons and provided images close-up of the largest Moons — Miranda, Ariel, Umbriel, Titania, and Oberon. Scientists were intrigued by Miranda's tortured surface. Titania is less than half the size of Earth's Moon.

A photograph taken in infrared light by the *Hubble Space Telescope* in 1995 showed the rings and different layers in Uranus' atmosphere. In 1998 the *Hubble Telescope* took an enhanced image of Uranus surrounded by four major rings, several Moons and clouds above the planet. In 2007, a Hubble image portrayed the Uranus ring system where the rings looked like spikes sticking out above and below the planet.

NEPTUNE

Neptune, which is a near twin of Uranus in size and density, is about one and a half times as far from the Sun as Uranus. Its atmosphere includes distinctive moving features with names like Dark Spot 2, Great Dark Spot, and the fast-moving Scooter. Winds can reach speeds of four hundred miles per hour. Neptune, thirty times farther from the Sun than Earth, is the most distant of the four giant planets in the Solar System.

Neptune takes nearly 165 years to orbit the Sun. It is nearly 2.8 billion miles from the Sun, has a diameter of 30,758 miles, and thirteen known

Neptune — Eighth Planet from the Sun. Neptune, named for the Roman God of the sea, is almost a twin to Uranus in terms of size and atmosphere. It has a diameter of 30,758 miles. Neptune's atmosphere appears blue because of its red-light-absorbing methane, which is present here in even greater quantity than on Uranus. The atmosphere of Neptune is mainly hydrogen and helium. The Great Dark Spot visible in the center of this photograph is a cyclonic feature similar to Jupiter's Great Red Spot and about the size of the Earth. This photograph is a composite of two images taken by *Voyager 2* in 1989 from a distance of 4.4 million miles.

Triton with Neptune in Background. Neptune floats like a blue billiard ball beyond its giant Moon, Triton, whose unique surface is seen through a very tenuous atmosphere. This portion of Triton, photographed by *Voyager 2*, shows this icy world's extraordinary surface that revealed a recent history of very unusual geologic activity. Triton orbits Neptune every four days and twenty-one hours.

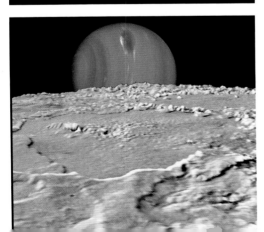

Neptune on Triton's Horizon. Triton is the largest of Neptune's eight known Moons. Triton reveals a tortured appearance caused by the surface disruptions that resulted from the energy released as a result of its capture into orbit around Neptune. The consequent heat generated in the Moon's interior undoubtedly fueled many vast volcanoes on Triton.

Moons. Triton, Neptune's largest Moon, has a diameter of 1,680 miles; above a surface coating of pink frost lays a thin atmosphere of nitrogen and methane. A close-up of Triton shows that it is mottled like a melon, with huge cracks and erupting geysers.

The *Voyager 2* spacecraft left Uranus behind in 1986 and three years later, in 1989, arrived at Neptune. On Earth, NASA linked together every large antenna possible to receive the faint and distant signals. Images from the space probe showed a bluish planet with a swirling story cloud structure and rings in the form of segmented arcs with thin connections that completed the circles. Neptune's largest Moon, Triton, seemed to have huge, spouting geysers or plumes of dark material erupting from its surface. *Voyager 2* discovered six of Neptune's Moons in 1989.

The Great Dark Spot, a large oval cloud about the same size as Earth was discovered by *Voyager 2* in 1989, but had vanished when the *Hubble Space Telescope* looked at Neptune in 1994. High winds were detected near the Great Dark Spot, making Neptune the windiest planet in the Solar System. *Voyager 2* also photographed a bright feature in Neptune's southern hemisphere called Scooter, made up of bright streaks of cloud that changed shape from day to day.

DWARF PLANET PLUTO

The discovery of Pluto in 1930 by Clyde Tombaugh, working at the Lowell Observatory in Flagstaff, Arizona, was hailed as the most important astronomical event in nearly one hundred years. Pluto, often called the "lost" planet, is the underdog of the Solar System. On August 24, 2006, the International Astronomical Union (IAU) held a vote to define the term "planet." Consequently, Pluto failed to make the grade and was re-designated a "dwarf planet." However, Pluto itself has not changed. It is still there orbiting the Sun and is as fascinating as ever.

The dwarf planet Pluto travels in an elliptical orbit around the Sun every 248 years. The mean distance of Pluto from the Sun is 3,631,000,000 miles. About half the size of Mercury, this dwarf planet is a small, icy world.

Pluto has three known Moons. In 1978 James W. Christy at the U.S. Naval Observatory discovered a Moon orbiting Pluto and named it Charon. Its diameter was estimated at 750 miles,

Getting Ready For Liftoff. Onlookers stand agape at the Atlas V rocket, poised to launch the *New Horizons* spacecraft (contained in the cone nose) on its nine-year journey to dwarf planet Pluto and beyond. The spacecraft was launched from Cape Canaveral Air Force Station on January 19, 2006. It sped away from Earth as the fastest spacecraft ever launched. It will reach dwarf planet Pluto in 2015.

more than half that of Pluto. In 2005, two tiny Moons with diameters of less than one hundred miles, named Nix and Hydra, were spotted on photographs taken by the *Hubble Space Telescope*.

The first Space probe ever to visit Pluto was already on its way when astronomers changed the status of the distant world. The *New Horizons* spacecraft blasted off from Cape Canaveral atop an Atlas V rocket on January 19, 2006 to start its nine-year journey to the distant, dwarf planet. When the last engine shut down, the spacecraft was traveling 36,373 miles per hour relative to Earth, a speed so fast that it reached the Earth's Moon orbit in only nine hours. It was the fastest traveling spacecraft ever sent anywhere.

New Horizons will reach Pluto, more than three billion miles from Cape Canaveral, on July 14, 2015. In February and March 2007, barely a year after launch, the *New Horizons* spacecraft conducted a Jupiter flyby and accomplished more than seven hundred scientific observations of Jupiter, its moons, its ring system, and its magnetosphere.

In 2010, the spacecraft passed the halfway point to its far-flung primary target, the dwarf planet Pluto. In spring 2011, New Horizons will pass the orbit of Uranus and begin the last, long leg of its journey — the almost billion-mile trip across Space between Uranus and Neptune. Once the spacecraft reaches Pluto, it will study the dwarf planet and its three Moons.

New Horizons is about the size of a piano and weighs, complete with fuel, just under 1,000 pounds. Inside this small package lie computing, guidance propulsion, communication, and power distribution systems and eight instruments. The instruments include two imagers, two spectrometers, two plasma/charged particle instruments, a radio-science package, and a dust counter to survey interplanetary space across the Solar System. These instruments will allow *New Horizons* to carry out not only an extensive investigation of Pluto, but also later possibly objects in the Kuiper Belt.

After the *New Horizons* speeds past Pluto in 2015, it will then go through the Kuiper Belt during the years 2016 and 2020, eventually leaving our Solar System to go toward the stars.

ASTEROIDS AND METEORITES

Asteroids

Billions of space rocks, known as asteroids, orbit the Sun within the inner Solar System. There are three main types of asteroids: those made of rock, those made of metal, and those that are a mixture of the two. Most of the asteroids, building blocks of the Solar System, tumble through

Asteroid Gaspra. Most asteroids are irregular in shape, like this close up of Gaspra. It was the first asteroid imaged at close quarters by the *Galileo* probe on October 29, 1991. About twelve miles long, it is one of the rocky asteroids. Note its numerous small crates and the dearth of moderate- or large-sized ones. Gaspra is not a primordial body but originates from the fragmenting of a much larger asteroid.

space in the vast gap between Mars and Jupiter. However, a few asteroids have orbits that bring them close to Earth.

Asteroids, more than three miles wide, probably strike Earth only once every ten million years. Rocks more than 160 feet wide may hit Earth once in a thousand years. Of more than 5,000 asteroids that approach Earth's orbit, fewer than a thousand are considered potentially hazardous objects that could threaten Earth. Since 1991, space probes have visited a number of bodies and observed them close-up.

Gaspra is a rock, the first to be seen close up. It orbits the Sun every three years and was photographed by the *Galileo* probe in October 1991. One 32-mile-long asteroid called Ida was found to have its own tiny Moon, named Dactyl. The *Galileo* spacecraft obtained images of the pair as it passed by them on August 28, 1993.

In June 1997, the *NEAR* (Near Earth Asteroid Rendezvous) probe showed that the asteroid Mathilde is riddled with giant craters and has an interior full of holes. In December 1998, the *NEAR* probe flew past asteroid Eros; the *Shoemaker* lander actually landed on Eros. In 2007, the *Dawn* probe was sent on its way to visit asteroids Vesta and Ceres in 2011 and 2015.

Meteorites

A meteorite is a piece of space rock that has survived a fall through Earth's atmosphere. Each year about 3,000 space rocks, too big to burn up in the Earth's atmosphere, land on the Earth's surface. Most fall into the ocean and are never found. Meteorites are divided into three types: iron meteorites, stony meteorites, and stony-iron meteorites. Every year about six meteorites are seen or heard falling to Earth.

Over a decade ago astronomers predicted that ice worlds smaller than planets would be found in a region named the Kuiper Belt. This region of Space was named for Dutch-American astronomer Gerard P. Kuiper, who, in 1951, championed the idea that the Solar System contains this distant family.

Kuiper and others envisioned a belt beyond planet Neptune and dwarf planet Pluto consisting residual material left over from the formation of planets. It seemed likely that these distant objects would be composed of water ice and various frozen gases — making them quite similar to the nuclei of comets.

In 1992, astronomers in Hawaii found the first of what is now a list of nearly a thousand icy bodies in that region. Some are quite large — hundreds of miles in diameter and are called planetoids.

Mike Brown, an astronomer at the California Institute of Technology, found many of the largest planetoids himself, but the one he found in November 2003, called Sedna, not only is bigger than any known Kuiper Belt object, other than dwarf planet Pluto, but also is many billions of miles farther from the Sun than the Kuiper Belt bodies. Sedna, at its closest approach to the Sun, is still two and a half times as far away as Pluto. Astronomer Brown suspects that there are larger worlds in our Solar System. He estimates that many other bodies the size of Sedna will eventually be found.

Earth's Meteor Crater. This large meteor crater, near Flagstaff, Arizona, is one of the youngest impact craters on Earth. It was excavated about 20,000 years ago when an iron mass struck flat-lying sedimentary rocks. It left a mile-wide bowl-shaped crater, surrounded by an extensive blanket of ejecta.

Meteorite on the Moon.
Most of the meteorites found on Earth are lumps of stone that were once part of an asteroid. However, a few meteorites came from the Moon, Mars or from comets. Meteorites are also found on Earth's Moon. Shown is *Apollo 17* astronaut-geologist Harrison "Jack" Schmitt investigating the site of a meteorite impact on the Moon during the final Apollo mission to the Moon in 1972. Schmitt's lunar rover, at right, weighed about 475 pounds on Earth but only seventy-nine pounds on the Moon with its low gravity. A meteorite found in Antarctica in 1981 was found to be part of the Moon, because it is almost identical to rocks brought back by *Apollo* astronauts.

THE FUTURE IN SPACE

THIS CHAPTER IS ABOUT THE FUTURE. NOT SOME DISTANT FUTURE THAT YOU can blissfully ignore, but which is imminent and whose progress can be plotted with some degree of precision. It is a future which is largely molded by a single, startling development in technology whose impact is just beginning to be felt — Space Travel.

What will the next ten, twenty, or thirty years bring? What new "technological breakthroughs" will occur? Naturally, no one can predict the future, but it is possible to make educated guesses about it and about the impact Space technology will have on our everyday lives. All indications are that the future holds many spectacular surprises about what is taking place in our Solar System.

The last half-century has been a period of pioneering. The Space Age has started, and is well under way: artificial satellites have been succeeded by Space Stations; unmanned probes have been sent to all the planets; Space telescopes are watching the Stars; and, of course, twelve men have walked on the Moon. So, what's next?

Mark Twain once said, "Predictions are very difficult to make, especially when they deal with the future." Keeping this in mind one might venture to identify some of the Space related activities that might occur in the near future.

WATER ON THE MOON

In 2009, forty years after the *Apollo 11* landing, lunar science again took the world by storm, upending long-held assumptions about our nearest heavenly neighbor — the Moon.

One of the most significant developments was the detection of water on the Moon's surface, previously thought to be dry. NASA's *Deep Impact* and *Cassini* spacecraft missions, along with India's lunar probe *Chandrayaan-1*, examined the Sun's infrared rays reflected off the Moon and detected the telltale fingerprints of water molecules. The amount of water was probably small — about an eyedropper-full for every two liters worth of lunar surface material.

NASA's *Lunar Reconnaissance Orbiter* (LRO) spacecraft was put into a Moon orbit in September 2009. From an altitude of just thirty-one miles, it will spend the next year imaging key parts of the lunar landscape with two narrow-angle cameras, which can see details as small as three feet wide.

A second water-hunting mission, the *Lunar Crater Observation and Sensing September* (LCROSS) was put in Moon orbit in September 2009. On October 9 NASA crashed a Centaur rocket into the South Pole's Cabeus crater while the LCROSS spacecraft followed in its wake, briefly examining the material that was kicked up before doing its own swan dive four minutes later. The 5,600 miles per hour impact carved out a hole sixty to one hundred feet wide and kicked up at least twenty-six gallons of water.

Lunar ice, if bountiful, not only gives future settlers something to drink, but could also be broken apart into oxygen and hydrogen. Both are valuable as rocket fuel, and the oxygen would give astronauts air to breathe.

A NASA DETOUR

America thought Space was worth a big effort in the 1960s. In 1961, President John F. Kennedy committed the nation to landing a man on the Moon by the end of the decade. This was achieved just in time with the *Apollo 11* mission in 1969. The historic images captured by the Apollo astronauts helped fuel our fascination with science and technology.

There are many excellent reasons for continued exploration of the Moon, which contains a future fusion-energy resource, helium 3, a rare (on Earth) helium isotope implanted by the solar wind. By-products of

Test Rocket Soars into Space. NASA's Ares I-X test rocket soars into the sky above its launch pad. The 327-foot-tall rocket produces 2.96-million pounds of thrust at liftoff and reaches a speed of 100-mph in eight seconds.

Ares I-X Test Rocket Launch. With more than twelve times the thrust produced by a Boeing 747 jet aircraft, NASA's Ares I-X test rocket roars off Launch Pad 39B at the Kennedy Space Center.

lunar helium 3 processing would include water, hydrogen, and oxygen — just what future explorers would need on a Moon-based colony, a base for travel to the rest of the Solar System. Lunar ice, if bountiful, could also give future settlers something to drink and could be broken apart into oxygen and hydrogen. Both are valuable as rocket fuel and the oxygen would give astronauts air to breath.

The escape velocity and the fuel required to launch from the Moon is 1/22 of the fuel needed for an Earth launch. The nuclear-ion-propulsion systems necessary for long travel to other planets would be much safer than launching boosters from Earth.

Perhaps the main purpose in returning humans to the Moon is to learn how to live on an alien world in preparation for a much more hazardous mission — the first human expedition to Mars. One of the most important objectives will be to see if astronauts can use lunar resources.

During the last decade, several countries have shown strong interest in continuing the exploration of the Moon. In early 2009, five spacecraft were orbiting the Moon: China's *Chang'e*, India's *Chandrayaan-1*, Japan's *Kaguya* and two small escorts, named *Okina* and *Ouna*. In June 2009, NASA's *Lunar Reconnaissance Orbiter* (LRO) joined these five spacecraft circling the Moon. China has announced its intention to launch a lunar rover in 2012, followed by a lunar sample-return mission in 2017. With a manned Space Station already on the drawing board, it seems only a matter of time before a Chinese citizen walks on the Moon.

During 2004-2009, NASA followed up the Space Exploration Vision announced by President George Bush in 2004 by preparing the transition from the Space Shuttle to a new Space architecture. The plan was to return to the Moon using a spacecraft that would hold six astronauts and a new launch vehicle to place the spacecraft in orbit. On October 28, 2009, NASA's 327-foot-tall *Ares I-X* test rocket lifted off Launch Pad 39B at the Kennedy Space Center on a test flight. As a result, it brought NASA one step closer to having a new launch vehicle for future Space exploration.

However, future plans to return American astronauts to the Moon took a detour in April 2010. President Barack Obama announced a new Space program for NASA and America that included designing a heavy-lift rocket that would take astronauts deep into Space by 2015, a visit to an asteroid by 2025, and then orbit Mars by the mid-2030s. Obama said in a speech at NASA's Kennedy Space Center, "I expect to live to see a landing on Mars." The president also said focusing on a manned Moon mission would be misguided. "We've been there before. There's a lot more of Space to explore and a lot more to learn when we do." Scientists are interested in studying asteroids, huge chunks of rock and gravel that orbit the Sun, because they could wipe out humanity if they collide with Earth.

Instead of returning to the Moon, Obama wants America to focus on research at the *International Space Station* and on helping private industry build commercial rockets to ferry people to the Space Station.

Journey to Mars

For many people, both inside and outside NASA, the dream of putting human beings on Mars has become the natural sequel to NASA's 1960s-1970s Moon expeditions.

Once man learns to live on the Moon, and the medical effects of long-duration flight in Space Stations have been assessed, a manned trip to Mars beckons as the next major goal. Such a journey would require major advances in space-propulsion systems as well as communication and life-support systems. Such a mission would represent a major breakthrough in the human exploration and settlement in Space. It would make the twenty-first century truly the "Age of Space." It would open the doors to all the wonders and dreams of the early space pioneers.

A round-trip to Mars is likely to take at least fifteen months. The first small steps on Mars will be a more complex, and more expensive, repeat of the first landing of men on the Moon.

Other nations are also planning to investigate Mars. The European Space Agency's (ESA) ExoMars mission, to be launched in 2013 or 2015, will deliver a large rover carrying a fully equipped laboratory able to analyze rock and soil samples for signs of life. Russia and China are planning *Phobos-Grunt*, a mission to land on the small Martian Moon Phobos.

All of these precursor missions will pave the way for the most ambitions robotic mission ever attempted—a Mars sample return.

TRIPS TO OTHER PLANETS

NASA has already sent the *New Horizons* spacecraft on its way to Pluto. The *Spirit* and *Opportunity* rovers are still studying rocks, soil, and minerals on Mars. The *Solar and Heliospheric Observatory* (SOHO) has been recording events on the Sun since 2008, and expects to observe furious bursts of activity there in 2012. Two space probes are set to circle Mercury in the near future. NASA has ambitious plans to return to Jupiter and, perhaps, place an orbiting spacecraft around its icy moon Europa.

Japan plans to launch *Venus C*, its first mission to Venus, in 2020. *Bepi-Colombo*, a joint mission with the European Space Agency, will follow in 2013.

The *Mars Reconnaissance Orbiter* will investigate Mars until December 31, 2010. In 2007 the *Dawn* probe was sent on its way to visit asteroids Vesta and Ceres in 2011 and 2015. Solar-powered *Juno* is scheduled to launch in 2011 on a five-year journey to orbit Jupiter. After arriving in 2016, it will probe deep into the planets atmosphere.

On October 22, 2008, India sent a space probe to the Moon and plans to launch its first astronaut sometime before 2015. If successful, it would become the fourth country to launch humans into orbit.

Future Russian space plans include sending a *Venera-D* space probe to Venus, slated for 2016; *Luna-Glob*, a lunar probe that is scheduled to launch in 2012; and *Phobos-Grunt,* to gather soil and rock samples from the Martian Moon Phobos.

The massive *Cassini-Huygens* spacecraft reached Jupiter in 2000. In December 2004, the Huygens lander, released from the *Cassini Orbiter*, plunged into the thick atmosphere of Saturn's Moon Titan, sending back enough data to keep scientists busy for decades. The scientific discoveries of the *Cassini-Huygens* are almost too numerous to describe. Among them are two new Moons of Saturn, spotted in June 2004.

SPACE-BASED ASTRONOMY

Space-based astronomy should continue to thrive. The remarkable *Hubble Space Telescope* should be able to continue examining the Universe until its replacement, the *James Webb Space Telescope* (JWST), arrives in orbit around 2013. Deep Space observatories such as JWST should be able to peer back in time to the so-called Dark Ages only a few hundred million years after the Big Bang.

Also under study, as part of the Beyond Einstein program, are two observatories, the *Laser Interferometer Space Antenna*, which will orbit the Sun measuring gravitational waves, and Constellation-X, which will observe matter falling into super-massive black holes.

On January 4, 2010, NASA's *Kepler Space Telescope* discovered five new planets orbiting distant Stars. The new planets, so-called hot Jupiters, are about the same mass as Jupiter and orbit very close to their host stars. Dubbed Kepler 4b, 5b, 6b, 7b and 8b, the five new planets range in temperature from 2,000 to 3,000 degrees Fahrenheit. The *Kepler Space Telescope,* launched from Cape Canaveral Air Force Station March 6, 2009, was designed to scour the Universe for evidence of planets with characteristics similar to those of Earth, so that one day Mankind might have a safe haven in 7.5-billion years time when our Sun explodes; that is if Mankind has not destroyed the planet himself long before that. Kepler will continue operations until at least November 2012. It is expected to take at least three years to locate and verify an Earth-size planet. The Kepler observations will tell us whether there are many stars with planets that could harbor life, or whether we might be alone in our galaxy.

FUTURE EARTH OBSERVATION

With global climate change at the top of political and scientific agenda, Earth observation will continue to be high priority throughout the world.

Europe is planning a series of Earth Explorer and Sentinel satellites as part of its Global Monitoring for Security program. The missions include a satellite dedicated to measuring the Earth's gravity field, a satellite to measure the thickness of surface ice sheets, a satellite to observe global wind profiles, and a satellite to study the interactions between clouds, solar radiation, and aerosols.

FUTURE COMMERCIAL
TRANSPORTATION SYSTEMS

On June 21, 2004, the first privately built spacecraft took off from an airport in Mojave, California, and reached an altitude of sixty-two miles, four hundred feet higher than the internationally agreed boundary that marks the beginning of Space. World famous aircraft designer Burt Rutan had turned his skills with lightweight composite fabrication to the problem of reaching Space, and created a remarkable new form of spacecraft called *SpaceShipOne*. Rutan's achievement in California demonstrated the possibility that suborbital Space flights is possible.

On December 7, 2009 billionaire Richard Branson's Virgin Galactic Company announced that commercial space travel will be available in the near future. The *Virgin Spaceship Enterprise*, a six-passenger, two-pilot spaceship, will take passengers on a two- and a half-hour suborbital trip for a fare of $200,000. People have already booked flights on one of Virgin's spaceships.

By 2014, engineer/pilot Mark Sirangelo and the Sierra Nevada's *Dream Chaser* spaceship could make its first orbital flight, and shortly thereafter hopes to be routinely ferrying up to seven people into low Earth orbit, perhaps on a visit to the *International Space Station*. The *Dream Chaser*, measuring about thirty feet long by twenty feet wide, is designed to hitch a ride atop a rocket such as the Atlas V. The Sierra Nevada Corporation, an electronic systems manufacturer in Sparks, Nevada, has more than twenty years experience building technologies ranging from handheld rockets to communication satellites.

The United States Air Force is also testing a new Space vehicle. In April 2010, it launched the *X-37B*, an unmanned, reusable Space plane. This 29-foot-long spaceship actually began as a possible next generation Space Shuttle, but NASA abandoned the project and the Air Force took control of it in 2006.

JOURNEYS TO THE STARS

What of destinations further away? Will mankind ever reach the Stars? In a way, interstellar travel has already begun, because the *Pioneer 10* and *11* and the *Voyager 1* and *2* spacecraft to the outer planets will eventually leave the Solar System and drift out into the Galaxy, carrying messages for other intelligent civilizations. Of course, the Pioneer and Voyager spacecraft instruments will have stopped working long before they leave the Solar System, but the spacecraft may one day be built that could survive a very long journey to the Stars. Hopefully some new interstellar propulsion systems would be designed to cut down travel time to the Stars.

In 1998, the *Deep Space 1* spacecraft set off to visit asteroid 1992KD and a comet. It traveled part of the way under ion engine power. Although the mission provided information about asteroids, its primary goal was to test the use of ion engines for travel in Deep Space. Standard rocket engines burn chemical fuels to propel spacecraft, but an ion engine uses electrically charged xenon gas, an element similar to neon. Ion engines use fuel ten times more efficiently than chemical engines, and, over time, achieve greater speeds.

FUTURE SPACE VEHICLES

NASA's Space Shuttle has certainly proved its value, but it is an expensive and imperfect vehicle for transporting cargo and passengers from Earth to orbit. It was indispensable for carrying and retrieving satellites in orbit and transporting supplies and personnel to the *International Space Station*. Eventually, though, NASA will develop new space vehicles.

An experimental space vehicle, the X-33 was one of the front-runners. This program was part of a cooperative venture among NASA, the United States Air Force, and private industry. The overall goal was to develop a simple, reusable Earth-to-orbit spacecraft that NASA could operate in much the same ways as conventional airlines.

The Venture Star (X-33) was designed to be the next generation space delivery system of the twenty-first century. Venture Star could revolutionize space transportation by delivering a wide range of payloads to Earth orbit more reliably and less expensively than any of today's launch vehicles. As a fully reusable, single-stage-to-orbit vehicle, Venture Star incorporated state-of-the-art technology and was built on lessons learned from older launch vehicles.

Venture Star (X-33) Space Vehicle.
This experimental Space plane was designed to carry people and cargo to and from Space in the twenty-first century.

RECOMMENDED READING

The following references offer additional information on the subjects covered in this introductory book.

Aldrin, Buzz and Ken Abraham. *Magnificent Desolation: The Long Journey Home From The Moon*. New York, New York: Harmony Books, 2009.

Alexander, George. *Moonport, U.S.A.* Publisher unknown.

Allen, Joseph P. *Entering Space: An Astronaut's Odyssey*. New York, New York: Stewart, Tabori & Chang, Publishers, 1985.

Allen, Lawrence J. *Man's Greatest Adventure*. Selah, Washington: BRM Selah Corporation, 1974.

Anderton, David A. *Man in Space*. Washington, D.C.: National Aeronautics and Space Administration, 1968.

Arnold, Wade. *Florida's Space Coast*. Charleston, South Carolina: Arcadia Publishing, 2009.

Baker, Wendy. *America in Space*. New York, New York: Crescent Books, 1986.

Barbree, Jay. *Live From Cape Canaveral*. New York, New York: Harper Collins, 2007.

Beasant, Pam. *1000 Facts About Space*. New York, New York: Kingfisher Books, 1992.

Beatty, J. Kelly, Carolyn Collins Petersen and Andrew Chaikin. *The New Solar System*. Cambridge, Massachusetts: Sky Publishing Corporation, 1999.

Bell, Jim. *Mars 3-D: A Rover's-Eye View of the Red Planet*. New York, New York: Sterling Publishing Company, 2008.

Bond, Peter. *Jane's Space Recognition Guide*. New York, New York: Harper Collins, 2008.

Bramwell, Martyn. *Book of Planet Earth*. New York, New York: Simon & Schuster, 1992.

Briggs, G. A. and F. W. Taylor. *The Cambridge Photographic Atlas of the Planets*. Cambridge, Massachusetts: Cambridge University Press, 1982.

Bryan, C. D. B. *The National Air and Space Museum*. New York, New York: Harry N. Abrams Inc., 1982.

Carlowicz, Michael. *The Moon*. New York, New York: Abrams, 2007.

Carr, Harriett. *Cape Canaveral: Cape of Storms and Wild Cane Fields*. St. Petersburg, Florida: Valkyrie Press, Inc., 1974.

Cernan, Eugene and Don Davis. *The Last Man on the Moon*. New York, New York: St. Martin's Press, 2009.

Chaikin Andrew. *A Man on the Moon: The Voyages of the Apollo Astronauts*. New York, New York: Penguin Group, 2007.

Chaikin, Andrew and Victoria Kohl. *Mission Control, This is Apollo: The Story of the First Voyages to the Moon*. New York, New York: Viking, 2009.
Voices from the Moon. London, England: Penguin Books Ltd., 2009.

Chaisson, Eric J. *The Hubble Wars*. New York, New York: Harper Collins, 1994.

Chartrand, Mark R. *Exploring Space: A Guide to Exploration of the Universe*. New York, New York: Western Publishing Company, 1991.

Clarke, Arthur C. *Man and Space*. New York, New York: Time-Life Books, 1969.

Collins, Michael. *Carrying the Fire: An Astronaut's Journeys*. New York, New York: Farrar, Straus and Giroux, 1974.

Cooper, Jr., Henry S. F. *Imaging Saturn: The Voyager Flights to Saturn*. New York, New York: Holt, Rinehart and Winston, 1982.

Coupler, Heather and Nigel Henbest. *Space Encyclopedia*. New York, New York: DK Publishing, Inc., 1999.

Cox, Donald W. *America's Explorers of Space*. Maplewood, New Jersey: Hammond Inc., 1967.

Darius, Jon, John Griffiths and Peter Turvey. *The Exploration of Space*. London, England: Science Museum, 1986.

Dickinson, Terence. *The Universe and Beyond*. Buffalo, New York: Firefly Books, 2004.

Dolan, Edward F. *Famous Firsts in Space*. New York, New York: E. P. Dutton, 1989.

Dyer, Alan. *Mission to the Moon*. New York, New York: Simon & Schuster Books, 2008.

Faherty, William Barnaby. *Florida's Space Coast: The Impact of NASA on the Sunshine State*. Gainesville, Florida: University Press of Florida, 2002.

Frazier, Kendrick. *Planet Earth Solar System*. Chicago, Illinois: Time-Life Books, 1985.

Goodwin, Simon. *Hubble's Universe: A Portrait of our Cosmos*. New York, New York: Penguin Books, 1997.

Gorn, Michael. *NASA: The Complete Illustrated History*. New York, New York: Merrell Publishers Ltd., 2005.

Grego, Peter. *The Great Big Book of Space*. New York, New York: Sandy Creek, 2009.

Hansen, James R. *First Man: The Life of Neil A. Armstrong*. New York, New York: Simon & Schuster, 2005.

Harris, Alan and Paul Weissman. *The Great Voyager Adventure*. Englewood Cliffs, New Jersey: Julian Messner, 1990.

Hartmann, William K., Ron Miller and Pamela Lee. *Out of the Cradle: Exploring the Frontiers Beyond Earth*. New York, New York: Workman Publishing Company, 1984.

Hockey, Thomas A. *The Book of the Moon: A Lunar Introduction to Astronomy, Geology, Space Physics, and Space Travel*. New York, New York: Prentice Hall Press, 1986.

Homan, Lynn M. and Thomas Reilly. *Historic Journeys Into Space*. Charleston, South Carolina: Arcadia Publishing, 2000.

Jenkins, Dennis R. and Jorge R. Frank. *The Apollo 11 Moon Landing*. North Branch, Minnesota: Speciality Press, 2009.

Jones, Tom. *Sky Walking: An Astronaut's Memoir*. New York, New York: Harper Collins, 2006.

Kerrod, Robin. *The Illustrated History of NASA*. New York, New York: Gallery Books, 1986.

Kerrod, Robin and Carole Stott. *Hubble: The Mirror on the Universe*. Buffalo, New York: Firefly Books Inc., 2007.

Knauer, Kelly, Editor. *Great Discoveries: Explorations that Changed History*. New York, New York: Time Books, 2009.

Kranz, Gene. *Failure is not an Option: Mission Control from Mercury to Apollo 13 and Beyond*. New York, New York: Simon & Schuster, 2000.

Kusky, Timothy. *Encyclopedia of Earth Science*. New York, New York: Facts On File, Inc., 2005.

LaFontaine, Bruce. *History of Space Exploration Coloring Book*. New York, New York: Dover Publications, Inc., 1989.

Langworthy, Fred H. *Thunder at Cape Canaveral*. New York, New York: Vantage Press, 1962.

Legault, Thierry and Serge Brunier. *New Atlas of the Moon*. Buffalo, New York: Firefly Books Inc., 2006.

Lethbridge, Cliff. *Cape Canaveral: 500 Years of History, 50 Years of Rocketry*. Merritt Island, Florida: SpaceCoast Cover Service, 2000.

Lipertito, Kenneth and Orville R. Butler. *A History of the Kennedy Space Center*. Gainesville, Florida: University Press of Florida, 2007.

Littmann, Mark. *Planets Beyond: Discovering the Outer Solar System*. New York, New York: John Wiley & Sons, 1988.

Macknight, Nigel. *Shuttle 2*. Osceola, Wisconsin: Motorbooks International, 1988.

Man, John, Editor. *The Encyclopedia of Space Travel and Astronomy*. New York, New York: Crescent Books, 1979.

Mason, Robert Grant, Editor. *Life in Space*. New York, New York: Time-Life Books Inc., 1983.

Mellberg, William F. *Moon Missions: Mankind's First Voyages to Another World*. Plymouth, Michigan: Plymouth Press, Ltd., 1997.

Miller, Ron and William K. Hartmann. *The Grand Tour: A Traveler's Guide to the Solar System*. New York, New York: Workman Publishing Company, 2005.

Mindell, David A. *Digital Apollo: Humans and Machines in Spaceflight*. Cambridge, Massachusetts: The MIT Press, 2008.

Moore, Sir Patrick. *Atlas of the Universe*. Buffalo, New York: Firefly Books, 2005.

Moore, Sir Patrick and H. J. P. Arnold. *Space: The First 50 Years*. New York, New York: Sterling Publishing Co., 2007.

Nelson, Craig. *Rocket Men: The Epic Story of the First Men on the Moon*. New York, New York: Penguin Group Inc., 2009.

Osborne, Ray. *Cape Canaveral*. Charleston, South Carolina Arcadia Publishing, 2008.

Parry, Dan. *Moon Shot: The Inside Story of Mankind's Greatest Adventure*. Ebury Press, 2009.

Pioneering the Space Frontier: The Report of the National Commission on Space. New York, New York: Bantam Books, 1986.

Powers, Robert M. *The World's First Spaceship: Shuttle*. Harrisburg, Pennsylvania: Stackpole Books, 1979.

Pyle, Rod. *Mission to the Moon*. New York, New York: Sterling Publishing Co., 2009.

Raeburn, Paul. *Uncovering the Secrets of the Red Planet*. Washington, D.C.: National Geographic Society, 1998.

Reynolds, David West. *Kennedy Space Center: Gateway to Space*. Buffalo, New York: Firefly Books Inc., 2006.

Ridpath, Ian. *Space*. Middlesex, England: Hamlyn Publishing, 1985.

Ridpath, Ian and Wil Tirion. *Stars & Planets*. Princeton, New Jersey: Princeton University Press, 2007.

Scarboro, C. W. *Pictorial History of Cape Kennedy 1950-1965*. Indialantic, Florida: South Brevard Beaches Chamber of Commerce, 1965.

Schulke, Flip, Debra Schulke, Penelope McPhee and Raymond McPhee. *Your Future in Space*. New York, New York: Crown Publishers, Inc., 1986.

Scott, Elaine and Margaret Miller. *Adventure in Space: The Flight to Fix the Hubble*. New York, New York: Hyperion Paperbacks, 1995.

Siepmann, H. R. and D. J. Shayler. *NASA Space Shuttle*. Runnymede, England: Ian Allan Ltd., 1987.

Smith, LeRoi. *We Came in Peace*. San Rafael, California: Classic Press Inc., 1969.

Smith, Melvyn. *An Illustrated History of Space Shuttle*. Newbury Park, California: Haynes Publications Inc., 1985.

Stroud, Rick. *The Book of the Moon*. New York, New York: Walker & Company, 2009.

Sutherland, Paul. *Where Did Pluto Go?* Pleasantville, New York: Reader's Digest, 2009.

Thompson, Neal. *Light This Candle: The Life & Times of Alan Shepard*. New York, New York: Three Rivers Press, 2004.

Torres, George. *Space Shuttle: A Quantum Leap*. Novato, California: Presidio Press, 1986.

Vanin, Gabriele. *A Photographic Tour of the Universe*. Buffalo, New York: Firefly Books Inc., 1999.

Watkins, Billy. *Apollo Moon Missions: The Unsung Heroes*. Westport, Connecticut: Greenwood Publishing Group, Inc., 2005.

Wilson, Andrew. *Space Shuttle Story*. New York, New York: Crescent Books, 1986.

Winchester, Jim, Editor. *Space Missions: The History of Space Flight*. San Diego, California: Thunder Bay Press, 2006.

Wolfe, Tom. *The Right Stuff*. New York, New York: Farrar, Straus and Giroux, 1979.

INDEX